Frontiers in Ceramic Science

(Volume 1)

(Functional Materials for Solid Oxide Fuel Cells: Processing, Microstructure and Performance)

Edited by

Moisés Rómolos Cesário

Unit of Environmental Chemistry and Interactions on Living - EA 4492
University of the Littoral Opal Coast (ULCO), France

&

Daniel Araújo de Macedo

Department of Materials Engineering, Federal University of Paraíba, Brazil

liability of Bentham Science Publishers shall be limited to the amount actually paid by you for the Work.

General:

1. Any dispute or claim arising out of or in connection with this License Agreement or the Work (including non-contractual disputes or claims) will be governed by and construed in accordance with the laws of the U.A.E. as applied in the Emirate of Dubai. Each party agrees that the courts of the Emirate of Dubai shall have exclusive jurisdiction to settle any dispute or claim arising out of or in connection with this License Agreement or the Work (including non-contractual disputes or claims).
2. Your rights under this License Agreement will automatically terminate without notice and without the need for a court order if at any point you breach any terms of this License Agreement. In no event will any delay or failure by Bentham Science Publishers in enforcing your compliance with this License Agreement constitute a waiver of any of its rights.
3. You acknowledge that you have read this License Agreement, and agree to be bound by its terms and conditions. To the extent that any other terms and conditions presented on any website of Bentham Science Publishers conflict with, or are inconsistent with, the terms and conditions set out in this License Agreement, you acknowledge that the terms and conditions set out in this License Agreement shall prevail.

Bentham Science Publishers Ltd.
Executive Suite Y - 2
PO Box 7917, Saif Zone
Sharjah, U.A.E.
Email: subscriptions@benthamscience.org

BENTHAM SCIENCE

CONTENTS

FOREWORD i

PREFACE ii

LIST OF CONTRIBUTORS iv

INTRODUCTION vi

CHAPTER 1 INTRODUCTION TO SOLID OXIDE FUEL CELLS 3
João Paulo de Freitas Grilo, Caroline Gomes Moura and Daniel Araújo de Macedo
   INTRODUCTION 3
   SOFC ELECTRODES 5
      Cathode 5
      Anode 6
   CONFLICT OF INTEREST 7
   ACKNOWLEDGEMENTS 7
   REFERENCES 7

CHAPTER 2 CATHODE MATERIALS FOR HIGH-PERFORMING SOLID OXIDE FUEL CELLS 9
Hanping Ding
   INTRODUCTION 10
      Technical Parameters for SOFC Cathodes 10
      Typical Materials for Cathode and their Conducting Nature 11
         Perovskite 11
         Mechanism of ORR Catalysis 13
         Prediction of Catalysts for ORR 14
         Basic Guidelines for Tailoring Electrical Properties 15
      Development of Low-temperature Cathode Materials 16
         The Parameters for Characterizing the Excellence of Cathodes 16
         (a). Electrical Conductivity 16
         (b). Polarization Resistance 17
         (c). Oxygen Surface Exchange and Diffusion Coefficient 17
         High-Performance Cathodes Operating at Low Temperatures 18
         (a). ABO3-type Perovskite 18
         (b). A2B2O5+δ-type Double Perovskite 19
   CONCLUDING REMARKS 21
   CONFLICT OF INTEREST 22
   ACKNOWLEDGEMENTS 22
   REFERENCES 22

CHAPTER 3 A BRIEF REVIEW ON ANODE MATERIALS AND REACTIONS MECHANISM IN SOLID OXIDE FUEL CELLS 26
Caroline Gomes Moura, João Paulo de Freitas Grilo, Rubens Maribondo do Nascimento and Daniel Araújo de Macedo
   INTRODUCTION 26
      Anode Materials 27
      NiO-GDC Composites 29
   ELECTROCHEMICAL PROCESSES IN SOFC ELECTRODES 33
   CONCLUDING REMARKS 37
   CONFLICT OF INTEREST 37
   ACKNOWLEDGEMENTS 37
   REFERENCES 38

CHAPTER 4 RECENT ADVANCES IN SYNTHESIS OF LANTHANUM SILICATE APATITE POWDERS AS NEW OXYGEN-ION CONDUCTOR FOR IT-SOFCS: A REVIEW 42

Chieko Yamagata, Daniel R. Elias, Agatha M. Misso and *Fernando S. Santos*

INTRODUCTION 43

LANTHANUM SILICATE APATITE (LSA) 45

LANTHANUM SILICATE APATITE (LSA) POWDER SYNTHESIS 48

Conventional Solid-State Reaction Synthesis 48

Sol-gel Synthesis 51

Coprecipitation Synthesis Method 55

Pechini Synthesis Method 56

Hydrothermal Synthesis Method 57

Freeze-Drying Synthesis 58

Plasma Spraying Synthesis 58

High-Energy Ball Milling Synthesis 59

Combined Synthesis Methods 60

CONFLICT OF INTEREST 63

ACKNOWLEDGEMENTS 63

REFERENCES 63

CHAPTER 5 A REVIEW ON SYNTHESIS METHODS OF FUNCTIONAL SOFC MATERIALS 70

Flávia de Medeiros Aquino, Patrícia Mendonça Pimentel and *Dulce Maria de Araújo Melo*

INTRODUCTION 70

TYPES OF SYNTHESIS METHODS 72

Solid State Reaction (Mixture of Oxides) 72

Experimental Procedure of Solid State Reaction 74

Combustion Synthesis 75

Experimental Procedure of the Combustion Synthesis 76

Sol-gel 78

Experimental Procedure of Sol-gel Method 78

Polymeric Complexing Method 79

Experimental Procedure of Polymeric Complexing Method 80

Co-precipitation Method 83

Experimental Procedure of Co-precipitation Method 84

CONFLICT OF INTEREST 85

ACKNOWLEDGEMENTS 85

REFERENCES 86

CHAPTER 6 CERAMIC HOLLOW FIBERS: FABRICATION AND APPLICATION ON MICRO SOLID OXIDE FUEL CELLS 88

Xiuxia Meng, Naitao Yang, Xiaoyao Tan and *Shaomin Liu*

INTRODUCTION 88

MICRO TUBULAR SOLID OXIDE FUEL CELL: ADVANTAGES AND FABRICATIONS 90

BASIC PRINCIPLES OF THE PHASE INVERSION TECHNIQUE 92

SINGLE-LAYER SPINNING FOR SOFC 94

DUAL-LAYER CO-SPINNING FOR MT-SOFC 98

TRIPLE-LAYER CO-SPINNING 101

CONCLUSION REMARKS 103

CONFLICT OF INTEREST 103

ACKNOWLEDGEMENTS 103

REFERENCES 103

CHAPTER 7 ELECTROLYTE HOLLOW FIBER AS SUPPORT VIA PHASE-INVERSION-BASED EXTRUSION/SINTERING TECHNIQUE FOR MICRO TUBULAR SOLID OXIDE FUEL CELL 107

Mohd Hafiz Dzarfan Othman, Siti Munira Jamil, Mukhlis A. Rahman, Juhana Jaafar and *A.F. Ismail*

INTRODUCTION 107

Phase-Inversion Based Extrusion/Sintering Technique for Fabrication of Ceramic Hollow Fiber 110

Electrolyte-supported Hollow Fiber 113

Electrolyte as Thin Layer 118

CONCLUDING REMARKS 126

CONFLICT OF INTEREST 127

ACKNOWLEDGEMENTS 127

REFERENCES 127

CHAPTER 8   PROTON CONDUCTING CERAMIC MATERIALS FOR INTERMEDIATE TEMPERATURE SOLID OXIDE FUEL CELLS 131

Narendar Nasani, Francisco Loureiro and *Duncan Paul Fagg*

INTRODUCTION 132

    Choice of Materials for PCFCs 135

ELECTROLYTES 135

    Synthesis 136

        Sintering and Microstructure 138

ANODES 143

    Cermet Compositions 145

        Synthesis, Phase Analysis and Microstructure 146

CATHODES 153

CONFLICT OF INTEREST 154

ACKNOWLEDGEMENTS 155

REFERENCES 155

SUBJECT INDEX 164

FOREWORD

This fascinating e-book clusters contributions from researchers who have dedicated the last years of their carrier to study materials, manufacturing processes and characterization techniques applied to the development of Solid Oxide Fuel Cells (SOFCs). These electrochemical devices that convert chemical energy into electricity are promising alternatives to traditional mobile and stationary power sources. Among their many advantages deserve special attention the high energy conversion efficiency and the excellent fuel flexibility. The development of high-performance functional SOFC is an important step towards reducing the operating temperature to 500 – 750 °C or lower. By doing this, the cell components can be easily and cost-efficiently produced. With this in mind, recent research around the world has focused on novel synthesis methods and processing routes to develop high performance components and single cells operating at reduced temperatures.

I am sure that this e-book reviews how processing conditions affect both microstructure and performance of functional SOFC materials.

Dr. Daniel Araújo de Macedo
Department of Materials Engineering
Federal University of Paraíba
Brazil

PREFACE

Solid Oxide Fuel Cells (SOFCs) are identified as a major technological promise for clean energy production. The development of functional materials for SOFC operating at intermediate temperatures (550 – 750 °C) requests not only a strict control of synthesis and processing conditions of ceramic/composite powders, but also a good understanding about the correlation between microstructure and electrochemical properties.

This e-Book aims to cluster contributions from the most productive and well-recognized researchers studying SOFC functional materials. Emphasis is on novel chemical/physical/mechanical processing routes towards the attainment of electrolyte and electrodes powdered/layered materials. Furthermore, the potential of the resulting microstructures toward SOFC applications has been checked using a combination of electron microscopy and electrical/electrochemical characterization techniques using symmetrical and/or single fuel cell configurations.

The book begins with an introductory chapter addressing the working principle of a SOFC and basic characteristics of SOFC electrodes. The second chapter is dedicated to cathode materials applied to intermediate and low-temperature SOFCs. The author proposes a comprehensive discussion on the cathode development, emphasizing its reaction mechanism, microstructural, characterization, and electrical performance. Studies of long-term chemical and mechanical stability have also been discussed.

The third chapter describes a review on anode materials, with focus on materials composition, synthesis methods, and electrical properties.

The forth chapter reports on the study of lanthanum silicate apatite based materials, drawing attention to their properties as electrolytes for SOFC. The authors propose a discussion on different synthetic methods to obtain apatite type electrolytes.

The fifth chapter presents a brief review on chemical/physical routes to prepare electrolyte and electrode materials for SOFC.

The sixth chapter reports on a recently phase inversion technique that is used to fabricate micro tubular solid oxide fuel cells (MT-SOFC). The authors propose a discussion on the development of this important manufacturing technique and their effects on the fuel cell performance.

The seventh chapter also discusses the use of the phase inversion based extrusion technique to fabricate MT-SOFC. Emphasis is given on the fabrication of electrolyte and how the fabrication parameters could affect the structure of the obtained electrolyte layer.

The eighth chapter reports on the study of proton conducting ceramic oxides with perovskite structure. The authors propose the development of electrolyte and electrode materials with combined properties of proton conductivity, high sinterability (in case of electrolytes), and chemical stability which make quite innovative research.

We would like to express our gratitude to all the eminent contributors for their excellent contributions and we believe that this e-book will be a reference to academic/industrial scientists from chemistry, physics, and materials science interested in the processing-microstructure-performance of SOFC materials.

Dr. Moisés Rómolos Cesário
Unit of Environmental Chemistry and Interactions on Living - EA 4492
University of the Littoral Opal Coast (ULCO)
France

&

Dr. Daniel Araújo de Macedo
Department of Materials Engineering
Federal University of Paraíba
Brazil

List of Contributors

Agatha M. Misso	Department of Materials Science and Technology, Nuclear and Energy Research Institute, Sao Paulo, Brazil
A. F. Ismail	Advanced Membrane Technology Research Centre, Universiti Teknologi Malaysia, Johor Bahru, Johor, Malaysia
Caroline G. Moura	Department of Mechanical Engineering, University of Minho, Braga, Portugal
Chieko Yamagata	Department of Materials Science and Technology, Nuclear and Energy Research Institute, Sao Paulo, Brazil
Daniel Araújo de Macedo	Department of Materials Engineering, Federal University of Paraíba, 58051-900, João Pessoa, Brazil
Daniel R. Elias	Department of Materials Science and Technology, Nuclear and Energy Research Institute, Sao Paulo, Brazil
Dulce M. de Araújo Melo	Department of Chemistry, Federal University of Rio Grande do Norte, Natal, Brazil
Duncan P. Fagg	Nanoengineering Research Group, Centre for Mechanical Technology and Automation, Department of Mechanical Engineering, University of Aveiro, Aveiro, Portugal
Fernando S. Santos	Department of Materials Science and Technology, Nuclear and Energy Research Institute, Sao Paulo, Brazil
Flávia de M. Aquino	Department of Renewable Energy Engineering, Federal University of Paraiba, João Pessoa, Brazil
Francisco Loureiro	Nanoengineering Research Group, Centre for Mechanical Technology and Automation, Department of Mechanical Engineering, University of Aveiro, Aveiro, Portugal
Hanping Ding	School of Petroleum Engineering, Xi'an Shiyou University, Xi'an, China. Colorado Fuel Cell Center, Department of Mechanical Engineering, Colorado School of Mines, Colorado, USA
João Paulo de F. Grilo	Department of Materials and Ceramic Engineering, University of Aveiro, Aveiro, Portugal
Juhana Jaafar	Advanced Membrane Technology Research Centre, Universiti Teknologi Malaysia, Johor Bahru, Johor, Malaysia
Mohd H. D. Othman	Advanced Membrane Technology Research Centre, Universiti Teknologi Malaysia, Johor Bahru, Johor, Malaysia
Mukhlis A. Rahman	Advanced Membrane Technology Research Centre, Universiti Teknologi Malaysia, Johor Bahru, Johor, Malaysia
Naitao Yang	School of Chemical Engineering, Shandong University of Technology, Zibo, China
Narendar Nasani	Nanoengineering Research Group, Centre for Mechanical Technology and Automation, Department of Mechanical Engineering, University of Aveiro, Aveiro, Portugal
Patrícia M. Pimentel	Federal Rural University of the Semi-Arid, Angicos, Brazil
Rubens M. do Nascimento	Department of Materials Engineering, Federal University of Rio Grande do Norte, Natal, Brazil

Shaomin Liu   Department of Chemical Engineering, Curtin University, Perth, Australia

Siti M. Jamil   Advanced Membrane Technology Research Centre, Universiti Teknologi Malaysia, Johor Bahru, Johor, Malaysia

Xiaoyao Tan   Department of Chemical Engineering, Tianjin Polytechnic University, Tianjin, China

Xiuxia Meng   School of Chemical Engineering, Shandong University of Technology, Zibo, China

INTRODUCTION

The Solid Oxide Fuel Cell (SOFC) technology has attracted significant attention due to the fuel flexibility and environmental advantages of this high efficient electrochemical device. However, typical SOFC operating temperatures near 1000 °C introduce a series of drawbacks related to electrode sintering and chemical reactivity between cell components. Aiming to solve these problems, researchers around the world have attempted to reduce the SOFC operating temperature to 500 – 750 °C or lower. It would result in the use of inexpensive interconnect materials, minimization of reactions between cell components, and, as a result, longer operational lifetime. Furthermore, decrease the operation temperature increases the system reliability and the possibility of using SOFCs for a wide variety of applications such as in residential and automotive devices. On the other hand, reduced operating temperatures contributes to increase ohmic losses and electrode polarization losses, decreasing the overall electrochemical performance of SOFC components. Thus, to attain acceptable performance, reducing the resistance of the electrolyte component and polarization losses of electrodes are two key points. Losses attributed to the electrolyte can be minimized by decreasing its thickness or using high conductivity materials such as doped ceria and apatite-like ceramics. Regarding electrode losses, the higher activation energy and lower reaction kinetics of the cathode compared with those of the anode, limits the overall cell performance. Therefore, the development of new functional SOFC materials with improved electrical/electrochemical properties combined with controlled microstructures become critical issues for the development of solid oxide fuel cells. These topics are systematic discussed along this e-book.

Frontiers in Ceramic Science

2

Frontiers in Ceramic Science

Volume # 1

Functional Materials for Solid Oxide Fuel Cells: Processing, Microstructure and Performance

Editors: Moisés R. Cesário and Daniel A. de Macedo

ISSN (Print): 2542-5250

eISSN (Online): 2542-5269

eISBN (Online): 978-1-68108-431-2

ISBN (Print): 978-1-68108-432-9

CHAPTER 1

Introduction to Solid Oxide Fuel Cells

João Paulo de Freitas Grilo[1], Caroline Gomes Moura[2] and Daniel Araújo de Macedo[3,*]

[1] *Department of Materials and Ceramic Engineering/CICECO, University of Aveiro, Aveiro, Portugal*

[2] *Department of Mechanical Engineering, University of Minho, Braga, Portugal*

[3] *Department of Materials Engineering, Federal University of Paraíba, 58051-900, João Pessoa, Brazil*

Abstract: Fuel cells are electrochemical devices that convert chemical energy into electrical energy with high potential for commercial power generation applications. Among various types of existing fuel cells, solid oxide fuel cell (SOFC) is one of the most promising types of fuel cells, due mainly to its ability to utilize several types of fuels such as hydrogen, CO, hydrocarbon fuels, and ethanol. This chapter introduces the reader into the fundamentals of SOFCs, including its working principle and the main components used as electrodes.

Keywords: Anodes, Cathodes, Ceramics, Chemical energy, Electrochemical, Electrodes, Electrolyte, Porous, Sealant, Solid Oxide Fuel Cells.

INTRODUCTION

Fuel cells are electrochemical devices that convert directly and efficiently chemical energy of a fuel gas into electrical energy. Furthermore, fuel cells are environmentally friendly devices whose efficiency is not limited to the Carnot-cycle and compared to others power generation systems with internal combustion, they do not produce significant amount of NO_x, SO_x, CO_x, and pollutants. The main fuel of fuel cells is hydrogen or hydrogen-rich fuels, this requirement makes fuel cell development a great challenge to researchers worldwide due to a number of problems involving hydrogen generation and storage. Fuel cells technology usage can be done in large stationary industrial plants as well as in vehicles and portable devices [1 - 5].

*Corresponding author Daniel A. de Macedo: Department of Materials Engineering, Federal University of Paraíba, 58051-900, João Pessoa, Brazil; Tel.: +55 83 3216-7076, Fax: +55 83 3216-7905; E-mail: damaced@gmail.com

Solid Oxide Fuel Cell (SOFC) is one of the most widely studied fuel cells, mainly because of their larger stability compared to other cell types, since it has a solid electrolyte, high efficiency and fuel flexibility. The main SOFC components are: porous cathode, porous anode, dense electrolyte, and sealants. The cathode is typically a solid state oxide which catalyzes oxygen reduction reaction, while anode is an oxide or cermet which catalyzes oxidation of a fuel, which can be either hydrogen or reformed hydrocarbons. The SOFC electrolyte must be an electronic insulating but ion-conducting material that allows only oxygen ions to pass through. Furthermore, this SOFC component must be dense to separate air and fuel, chemically and structurally stable over a wide range of partial pressures of oxygen and temperatures. Sealant materials, often used during the manufacture of single SOFCs, should provide a viscous behavior for coupling the compensating tolerances and other materials avoiding failures, which guarantees a hermetic seal.

Usually a SOFC operates at high temperatures in a range of 600 - 1000 °C allowing internal reforming of fuel. The characteristics of high operating temperature of SOFC present great challenges related to the cell lifetime and materials degradation. Therefore, there is a great interest in reducing the SOFC operating temperature to a range of 500 – 800 °C or lower, which reduces their production costs as well as stability and degradation issues. The operating principle of SOFC is schematically illustrated in Fig. (**1**). The fuel, hydrogen or a hydrocarbon gas, permeates into the anode compartment and the oxygen, from the air, into the cathode. At the anode (fuel electrode side), fuel is oxidized according to the reaction (Eq. 1):

$$H_{2(g)} + O^{2-} \rightarrow H_2O_{(g)} + 2e^- \tag{1}$$

The electrons are transported to the cathode through an external circuit. At the cathode the oxygen is reduced with the incoming electrons from external load according to the reaction (Eq. 2):

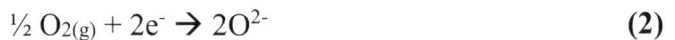

$$\tfrac{1}{2} O_{2(g)} + 2e^- \rightarrow 2O^{2-} \tag{2}$$

Generated oxygen ions migrate to anode across the electrolyte, hence, the fuel is oxidized by incoming oxygen ions. Therefore, this electrical connection allows a continuous supply of oxygen ions from the cathode to the anode, whilst maintaining an overall balance of electrical charge, thus producing electrical energy. The products of these reactions (Eq. 1 and 2) are only water and heat (Fig. **1**). Most of the electrochemical reactions in a cell occur in the so-called triple phase boundary (TPB), which is the contact region between gas phase, electrode and electrolyte [2, 6 - 8].

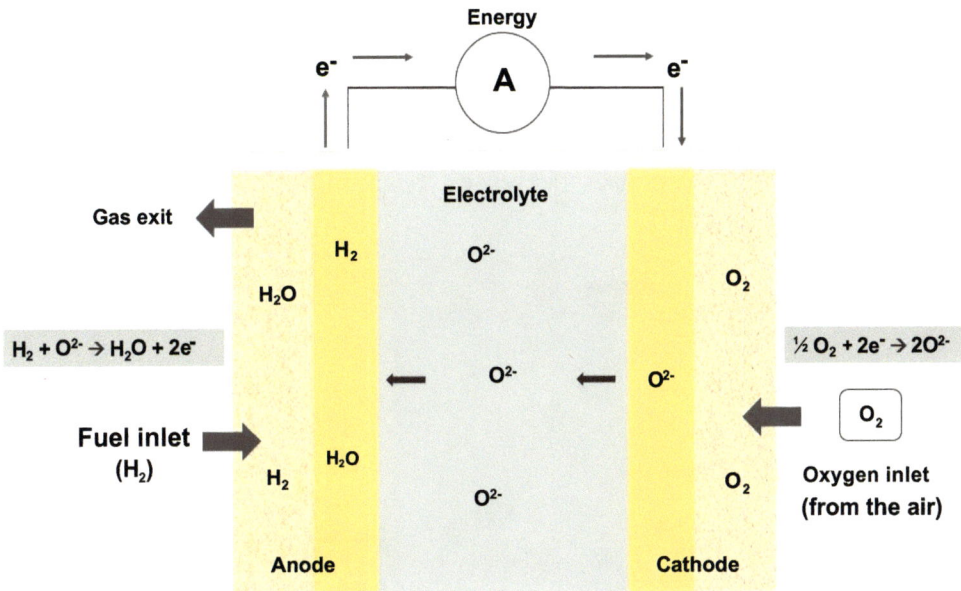

Fig. (1). The working principle of a SOFC.

Besides showing the working principle of a typical SOFC, this chapter is also focused on a brief review on the main SOFC electrodes (cathode and anode). Materials, processing and obtaining methods of electrodes and electrolyte materials will be discussed in the following chapters.

SOFC ELECTRODES

Cathode

The cathodes for SOFC must have several properties, including: (a) high electrical conductivity; (b) high catalytic activity for the oxygen reduction; (c) good compatibility with others cell components; (d) suitable porosity (approximately 30-40%); (f) thermal expansion coefficient matching those of other components; (g) chemical stability during fabrication and operation; (h) low manufacturing cost and (i) extensive TPB (triple phase boundary). In the early development of SOFC, platinum was used as cathode, nevertheless it seemed very costly for commercial use. The cathode has the function to reduce the oxygen molecule, transport ion to the electrolyte and provides electrical current resulting from reduction reaction of oxygen. Thus, the choice of electrode material is important for high performance of the cell and, thereby, avoids undesirable chemical reactions [9 - 12]. In the cathode, the reaction is shown in Eq. 3:

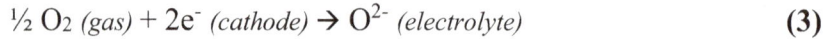

$$\frac{1}{2} O_2 \ (gas) + 2e^- \ (cathode) \rightarrow O^{2-} \ (electrolyte) \tag{3}$$

The TPB area is schematically illustrated in Fig. (**2**). Any disruption in connectivity among these phases decreases the number of points to occur the electrochemical reaction [13].

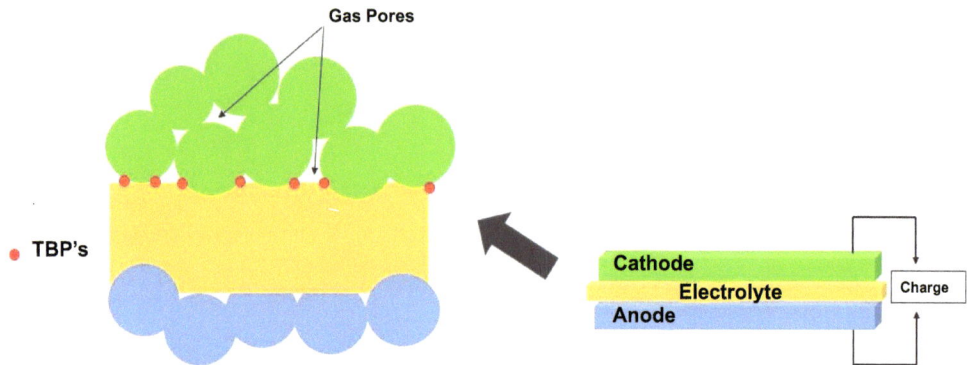

Fig. (2). Schematic representation of triple phase boundaries.

Anode

The anode is the electrode where the fuel oxidation occurs. As the cathode, this component must also exhibit high electronic conductivity, good catalytic activity for the fuel oxidation reactions and sufficient porosity to allow the transport of fuel to the anode/electrolyte interface and the removal of reaction products. In addition, the anode should be chemically stable and thermally compatible with the other SOFC components [1, 14].

The electrochemical performance of the anode depends on the charge transport resistance (electrons and ions), inside the anode and the anode/electrolyte interface, and the resistance of gas transport. The increase of the triple phase boundaries (TPB) length, by microstructural optimization and phase composition, are the most efficient ways to improve the electrochemical performance of anodes [15, 16].

Internal reform and tolerance to sulfur-containing compounds are also essential to the anodes, especially when a hydrocarbon fuel is used, *e.g.* methane. The porosity of the anodes is a very important factor, not only because it is related to high densities of triple phase boundaries, but also because it avoids mass transport limitation. In this regard, many studies have reported the use of pore formers (graphite, starch, citric acid, *etc.*) in order to obtain suitable porosity in anodes [17 - 19]. However, due to the tendency to agglomerate of pore-forming agents, it is sometimes difficult to ensure good structural performance and permeation of gases in these electrodes [20].

CONFLICT OF INTEREST

The authors confirm that they have no conflict of interest to declare for this publication.

ACKNOWLEDGEMENTS

Declared none.

REFERENCES

[1]   Atkinson, A.; Barnett, S.; Gorte, R.J.; Irvine, J.T.; McEvoy, A.J.; Mogensen, M.; Singhal, S.C.; Vohs, J. Advanced anodes for high-temperature fuel cells. *Nat. Mater.,* **2004**, *3*(1), 17-27.
[http://dx.doi.org/10.1038/nmat1040] [PMID: 14704781]

[2]   Brett, D.J.; Atkinson, A.; Brandon, N.P.; Skinner, S.J. Intermediate temperature solid oxide fuel cells. *Chem. Soc. Rev.,* **2008**, *37*(8), 1568-1578.
[http://dx.doi.org/10.1039/b612060c] [PMID: 18648682]

[3]   Sun, C.; Stimming, U. Recent anode advances in solid oxide fuel cells. *J. Power Sources,* **2007**, *171*, 247-260.
[http://dx.doi.org/10.1016/j.jpowsour.2007.06.086]

[4]   Murray, E.P.; Tsai, T.; Barnett, S.A. A direct-methane fuel cell with a ceria-based anode. *Nature,* **1999**, *400*, 649-651.
[http://dx.doi.org/10.1038/21781]

[5]   McIntosh, S.; Gorte, R.J. Direct hydrocarbon solid oxide fuel cells. *Chem. Rev.,* **2004**, *104*(10), 4845-4865.
[http://dx.doi.org/10.1021/cr020725g] [PMID: 15669170]

[6]   Carrette, L; Friedrich, A; Stimming, U. Fuel Cells - Fundamentals and Applications. *Wiley online Libr.,* **2001**, *1*, 5-39.

[7]   Stambouli, A.B.; Traversa, E. Solid oxide fuel cells (SOFCs): A review of an environmentally clean and efficient source of energy. *Renew. Sustain. Energy Rev.,* **2002**, *6*, 433-455.
[http://dx.doi.org/10.1016/S1364-0321(02)00014-X]

[8]   Taroco, H.A.; Santos, J.A.; Domingues, R.Z.; Matencio, T. Ceramic Materials for Solid Oxide Fuel Cells. *Adv Ceram Synth Charact Process Specif Appl,* **2011**, *5*, 423-446.

[9]   Yokokawa, H.; Horita, T. Cathodes. In: *High temperature solid oxide fuel cells: fundamentals, design and applications*; Singhal, C.S.; Kendall, K., Eds.; Elsevier Science, **2003**; pp. 119-147.
[http://dx.doi.org/10.1016/B978-185617387-2/50022-2]

[10]   Jiang, S.P.; Jian, L. Cathodes. In: *Solid oxide fuel cells: materials properties and performance*; Fergus, J.W.; Hui, R.; Li, X., Eds.; CRC Press, **2009**; pp. 131-171.

[11] Sun, C.; Hui, R.; Roller, J. Cathode materials for solid oxide fuel cells: a review. *J. Solid State Electrochem.,* **2010**, *14*, 1125-1144.
[http://dx.doi.org/10.1007/s10008-009-0932-0]

[12] Steele, B.C.; Bae, J.M. Properties of $La_{0.6}Sr_{0.4}Co_{0.2}Fe_{0.8}O_{3-x}$ (LSCF) double layer cathodes on gadolinium-doped cerium oxide (CGO) electrolytes. *Solid State Ion.,* **1998**, *106*, 255-261.
[http://dx.doi.org/10.1016/S0167-2738(97)00430-X]

[13] Adler, S.B. Factors governing oxygen reduction in solid oxide fuel cell cathodes. *Chem. Rev.,* **2004**, *104*(10), 4791-4843.
[http://dx.doi.org/10.1021/cr020724o] [PMID: 15669169]

[14] Liu, J.; Barnett, S.A. Operation of anode-supported solid oxide fuel cells on methane and natural gas. *Solid State Ion.,* **2003**, *158*, 11-16.
[http://dx.doi.org/10.1016/S0167-2738(02)00769-5]

[15] Holzer, L.; Münch, B.; Iwanschitz, B.; Cantoni, M.; Hocker, T.; Graule, T. Quantitative relationships between composition, particle size, triple phase boundary length and surface area in nickel-cermet anodes for Solid Oxide Fuel Cells. *J. Power Sources,* **2011**, *196*(17), 7076-7089.
[http://dx.doi.org/10.1016/j.jpowsour.2010.08.006]

[16] Mizusaki, J.; Tagawa, H.; Saito, T.; Yamamura, T.; Kamitani, K.; Hirano, K. Kinetic studies of the reaction at the nickel pattern electrode on YSZ in H_2-H_2O atmospheres. *Solid State Ion.,* **1994**, *70-71*, 52-58.
[http://dx.doi.org/10.1016/0167-2738(94)90286-0]

[17] Zhu, W.Z.; Deevi, S.C. A review on the status of anode materials for solid oxide fuel Cells. *Mater. Sci. Eng.,* **2003**, *362*, 228-239.
[http://dx.doi.org/10.1016/S0921-5093(03)00620-8]

[18] Haslam, J.J.; Pham, A.Q.; Chung, B.; Dicarlo, J.F.; Glass, R.T. Effects of the Use of Pore Formers on Performance of an Anode Supported Solid Oxide Fuel Cell. *J. Am. Ceram. Soc.,* **2005**, *88*, 513-518.
[http://dx.doi.org/10.1111/j.1551-2916.2005.00097.x]

[19] Clemmer, R.M.; Corbin, S.F. Influence of porous composite microstructure on the processing and properties of solid oxide fuel cell anodes. *Solid State Ion.,* **2004**, *166*, 251-259.
[http://dx.doi.org/10.1016/j.ssi.2003.12.009]

[20] Hu, J.; Lü, Z.; Chen, K.; Huang, X.; Ai, N.; Du, X.; Fu, C.; Wang, J.; Su, W. Effect of composite pore-former on the fabrication and performance of anode-supported membranes for SOFCs. *J. Membr. Sci.,* **2008**, *318*, 445-451.
[http://dx.doi.org/10.1016/j.memsci.2008.03.008]

Frontiers in Ceramic Science, 2017, *Vol. 1*, 9-25   9

CHAPTER 2

Cathode Materials for High-Performing Solid Oxide Fuel Cells

Hanping Ding[1,2,*]

[1] *School of Petroleum Engineering, Xi'an Shiyou University, Xi'an 710065, China*

[2] *Colorado Fuel Cell Center, Department of Mechanical Engineering, Colorado School of Mines, Colorado, 80401, USA*

Abstract: It is well recognized that the development of low-temperature solid oxide fuel cells (LT-SOFCs) replies on the exploration of new functional materials and optimized microstructures with facilitated oxygen reduction reaction (ORR) that involves complicated electrochemical processes occurring at triple-phase boundaries (TPB). This urgent and critical demand promotes great research efforts on pursuing superior catalysts as electrodes owing comprehensive electrochemical and physicochemical properties, and relevant catalyst optimization on materials and microstructures. The material development is mostly based on perovskite with extensive doping strategies to maximize the catalytic activity while other properties such as stability, thermal and chemical compatibility, *etc.* are well compromised. Other types of materials such as K_2NiF_4, double perovskite were also studied as potential candidates, owing to the excellence of catalytic activity resulting from the special features of crystal structures. In this chapter, the fundamental knowledge of cathode is briefly introduced, such as reaction processes of ORR, catalysis mechanism and defect transport. Several typical perovskites are reviewed to better understand the required specific material properties for an excellent ORR catalyst as cathode material that can be operated at practical low temperatures (350~500 °C). Particularly, recent development of the layered perovskites is specifically introduced because they show very promising performance at low temperatures due to the fast oxygen exchange and oxygen diffusion yielded by the ordered cation distribution in crystal.

Keywords: Catalysis mechanism, Catalytic activity, Cathode, Layered perovskite, Oxygen diffusion, Oxygen reduction reaction, Oxygen surface exchange, Perovskite, Solid oxide fuel cells, Three-phase boundary.

---

* **Corresponding author Haping Ding:** Colorado Fuel Cell Center, Department of Mechanical Engineering, Colorado School of Mines, Colorado, 80401, USA; Tel: 1-303-384-2096; E-mail: hding@mines.edu; hanpingding@gmail.com

Moisés R. Cesário & Daniel A. de Macedo (Eds.)

INTRODUCTION

In this chapter, the recent development of cathode materials, which are operated at low operating temperature (350 ~ 600 °C) is discussed. With emphasis on how the candidate materials are selected as potential low-temperature operating cathodes, mechanism of oxygen reduction reaction, criteria of promising cathode and role of mixed ionic-electronic conductors are also discussed.

Technical Parameters for SOFC Cathodes

As an air electrode, the oxygen reduction reaction (ORR) occurs at the three phase boundaries where the oxygen gas, electrolyte and cathode surfaces meet. The produced negatively charged oxygen ions transfer through the electrolyte conducting membrane and then react with hydrogen molecules to form water while the electrons released from fuel of hydrogen have to pass the external circuit to form the current. Therefore, the ORR is a very critical step to determine the initial kinetics of total reaction. The oxygen reduction on the cathode surface is believed to include several sub-steps which separately determine the limiting step, such as oxygen absorption, charged, dissociation and desorption, *etc.* (Fig. **1**) shows the reaction steps at TPB area (pure electronic conductor is used to illustrate for simplicity). The oxygen molecules are absorbed on the surface or the TPB sites first and then move towards TPB area to be dissociated, where oxygen ions are formed through electrochemically charged by the electrons. Consequently, the oxygen ions should have to leave the sites and move towards electrolyte and incorporate into it. If a mixed ionic and electronic conductor is used, the places for oxygen dissociation can be extended to the whole cathode surface. Therefore, the oxygen ions can reach the electrolyte membrane by another pathway of bulk cathode. The reaction kinetics can be significantly increased by this extension of reaction sites and diffusion paths.

In order to facilitate ORRs to proceed fast, several technical requirements have to be satisfied. (a) Electrical conductivity: since ORR is an electrochemical reaction, a certain conductivity is needed to allow electron conduction. The mixed conductor of electrons and oxygen ions is preferred because the more active sites are created; (b) Catalytic activity: the property of surface chemistry need to allow the absorption and desorption of various oxygen-related species on the cathode surface; (c) porosity: the gas diffusion from the layer surface to the cathode/electrolyte interface should be fast to minimize the concentration over-potential; (d) thermal and chemical compatibility with electrolyte: the thermal expansion of cathode bulk should be close to that of electrolyte to avoid the potential delamination between two layers and to increase the resistance against thermal shocks. The cathode should not react with electrolyte to form any insulating phases that slower down or block the further proceeding of oxygen reduction; (e) chemical stability: the cathode must be chemically stable in some case of low oxygen partial pressures when large amount of oxygen is consumed by reactions, causing the oxygen absent at electrolyte/cathode interface area. Overall, the cathode performance is determined by all of these physical, chemical, or electrochemical parameters; one bad-done aspect can deteriorate the cathode behavior in operation.

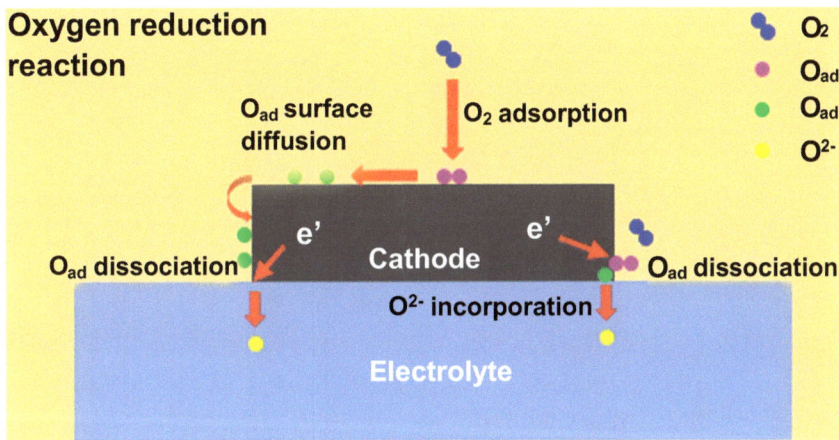

Fig. (1). The schematic of catalysis mechanism for ORR on the cathode surface (pure electronic conductor).

Typical Materials for Cathode and their Conducting Nature

Perovskite

Searching for a conductive oxide which can sustain good relevant properties after the processes of material synthesis and fuel-cell fabrication conditions. Many oxides with perovskite structure are ideal material candidates, which are still in great interest in current activities. Perovskite is a class of compounds which have the same type of crystal structure as $CaTiO_3$ ($A^{2+}B^{4+}X^{2-}_3$). Due to the high tolerance factor of the crystal structure, perovskite offers wide flexibility for improving the properties of materials, such as catalytic activity, electronic or ionic conductivity, chemical stability and thermal behavior, *etc* [1]. Many useful properties of the perovskite oxides are primarily determined by the B-site cations while they can be also tuned by A-site cations. In this perovskite structure as shown in Fig. (**2**), oxides typically adopt a cubic structure (sometimes it also distorts to other crystal structure depending on the atoms and preparation conditions).

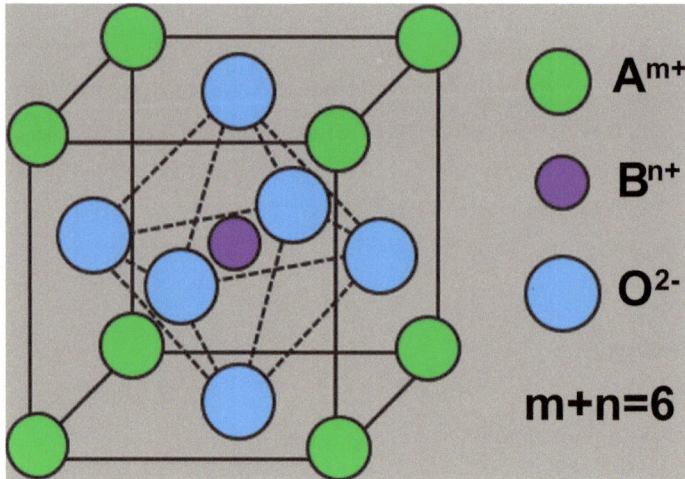

Fig. (2). The unit-cell structure of ABO_3-type perovskite. The A-site cation occupies the larger spaces of 12-fold oxygen coordinated interstitials; B-site cation occupies the smaller octahedral holes (6-fold coordination).

A large alkaline earth or rare earth cation occupies A-site locating at the corners of the cube, while smaller transition metal element sits at the body center, and oxygen atoms at the centers of six cubic faces. The A-site cations (such as La, Ca, Sr, Ba, Pr and Sm, *etc.*) with lower valence at the interstitial sites are surrounded by four octahedron and coordinated to twelve oxygen anions while the B cations (such as Cr, Ti, Fe, Cu, Ni, Ce, Y, Yb and Zr, *etc.*) at the center of octahedron are coordinated to six oxygen atoms at the corners.

Perovskite oxides generally involve the structural distortion such as the tilting of the octahedron and displacement of cations at either A or B site, resulting from equilibrium states of charged cations and anions with different ionic sizes and valences. While this distortion may achieve some required electrical properties, it also leads to the structural instability. In order to quantify the degree of this distortion, a term called Goldschmidt tolerance factor (t) is used to describe it, according to the equation (1) [2]:

$$t = \frac{r_A + r_O}{\sqrt{2}(r_B + r_O)} \tag{1}$$

When the so-defined tolerance factor is in the range of 0.77~1.0, the crystal structure can be stable. With t closer to unity, higher symmetry and smaller volume of unit cell can be obtained by the proper choice of elements at the cation sites. When A-site cation is replaced by larger cation, tolerance factor increases;

and when it is replaced by smaller cation, it decreases. On the other hand, perovskites can accommodate element replacement in a wide range due to the flexibility of crystal structure, which usually serves as a basis for tailoring specific properties, such as introducing large amount of oxygen vacancies or small fraction of cation deficiency. Some perovskite-type oxides for cathode are doped lanthanum manganite ($LaMnO_3$) [3], lanthanum cobaltite ($LaCoO_3$) [4] and barium cobaltite ($BaCoO_3$) [5].

Mechanism of ORR Catalysis

There have been some experimental and modelling studies contributed to the better understanding of the mechanisms occurring at SOFC cathodes interfaces. These studies have provided us some valuable information about the relationship between alterable material structure and cathode polarization resistance. It is generally believed that SOFC cathode materials should favorably be mixed ionic-electronic conductors, *i.e.* electron and oxygen-ion conduction (MIEC) at the same time. As shown in Fig. (**3**), when cathode is a pure electronic conductor the ORR is restricted only to take place at the TPBs of gas phase, electrolyte and cathode [6]. There is a possible mechanism for ORR at MIEC: the oxygen specie is firstly adsorbed at the cathode surface. The formed oxygen species diffuse along cathode surface to the TPB area, and then become discharged and incorporated into oxide-ion conducting electrolyte. If cathode is a MIEC, the reaction kinetics of the ORR could be enhanced significantly (Fig. **3b**). Because MIEC can provide more active sites for oxygen reduction at both electrode surface and bulk which allow oxygen ion to diffuse in these two different pathways. Therefore, two parallel pathways are available to have oxygen ions incorporating into the electrolyte: the surface and the bulk. There are two terms used for describing the above steps: oxygen surface exchange coefficient (k) for electrochemical adsorption and charging at electrode surface and the oxygen diffusion coefficient (D) for oxygen ion diffusion in the electrolyte. The actual extension of the electrochemical sites for the ORR is closely associated with ratio of surface exchange to ionic conductivity in relation to particle size (L), Lk^c/D^c (c represents a constant deduced from experiments). The smaller ratio indicates more area of the MIEC surface available for oxygen reduction.

The basic process for a catalyst to catalyze the ORR reaction can be simply described in this way. According to Sabatier principle, the oxygen species in the whole reaction process should be absorbed and desorbed for a right time. In the other words, the interactions between the catalyst and oxygen species should be just right: neither too strong nor too weak. If the bonding is too weak, the oxygen species fails to bind to catalyst and no reaction takes place. On the other hand, if the bonding is too strong, the surface of catalyst can be blocked by the reactants or products that would fail to dissociate. Therefore, the oxygen should be easily absorbed and also quickly desorbed very fast on the cathode surface.

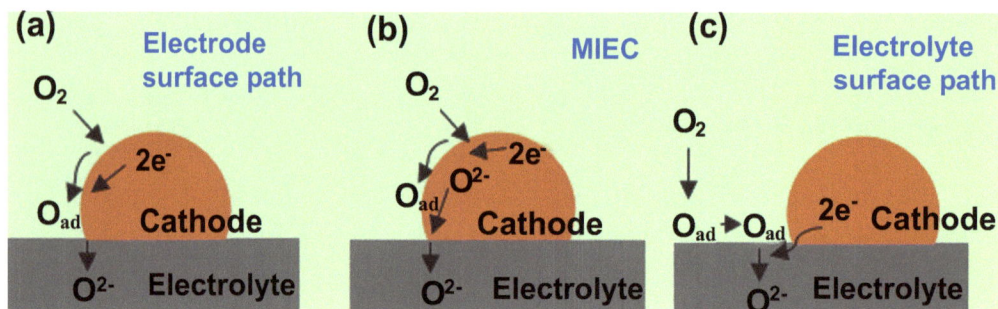

Fig. (3). Schematic of the three different mechanisms for oxygen reduction and pathways of oxygen ions.

Prediction of Catalysts for ORR

In order to identify the right catalysts for ORR, a design principle that links material properties to the catalytic activity should be proposed. Recent ab initio [7] and experimental studies [8, 9] have identified a unique catalyst property (or activity descriptor) that governs the strength of the metal-oxygen bonding strength and ORR activity of platinum-based metals in acid. Lately, Yang *et al.* demonstrated that the catalytic activity for transition metal oxides is mainly correlated to e_g orbital occupation (the electron orbital due to splitting of d-orbitals in the crystal field) and the extent of covalence between transition metal and oxygen ions, which is regarded as a secondary activity descriptor [10]. Interestingly, it is found that an M-shaped relationship between catalytic activity and d-orbital electron yields maximum activity attained near d^4 and d^7. The intrinsic ORR activity of the perovskite oxides as a function of the number of electrons filled at e_g orbital exhibits a volcano shape. Why e_g filling is closely related to ORR activity? For example, too little e_g-filling in $LaCrO_3$ ($t_{2g}^3 e_g^0$ for $x=0$, $t_{2g}^{2.5} e_g^0$ for $x=0.5$) can result in $B-O_2$ bonding that is too strong, whereas too much e_g-filling in $LaFeO_3$ and $LaNiO_3$ can lead to an O_2 interaction that is too weak. Neither situation is optimum for ORR activity. On the other hand, a moderate amount of e_g-filling (~1) in $LaMnO_3$, $LaCoO_3$ and $LaNiO_3$ yields the highest activity. So e_g-filling represents a primary descriptor.

Basic Guidelines for Tailoring Electrical Properties

Electrochemical properties of perovskites originate from the cations at B site and can be tuned by cations at A site. For $LaMnO_3$ or $LaCoO_3$ as an example, the trivalent lanthanum at A-site is often replaced by divalent Sr. According to electro neutrality law, the effective negative charge induced by element replacement is compensated either by an increase in valence of B-site cations to form a mixed valence state (Mn^{3+}/Mn^{4+}) (so-called "electronic compensation) or formation of oxygen vacancies (so-called "ionic compensation"). For electronic conduction, in many cases, particularly for cathode materials, small polaron hopping might be the important mechanism. The electronic conduction can be attributed to the electrons occupying hydrogenic orbitals with wavefunction are localized by potential fluctuations [11]. The simple scene is that electrons or holes are transferred along B-O-B bonds in the three-dimensional BO_6 octahedra network through the heterovalent cations at the B site. The covalency of B-O chemical bond is closely associated with the electrical properties.

Fig. (**4**) shows the crossing orbitals of metal cation at B site and adjacent oxygen atom, which is responsible for the hopping of electron [1]. As can be seen, the p_σ orbitals of oxygen atom are attracted by the nuclear charge of the B cation and then combined tightly with the p_σ orbitals of B cation. This colinear orbital overlap can strongly screen the t_{2g} orbitals which spread towards p_π orbitals of oxygen atom. Thereby, the d-orbital electrons of t_{2g} can drift into the t_{2g} orbital of a neighboring B-site cation by p_π orbital of oxygen atom between two B cations.

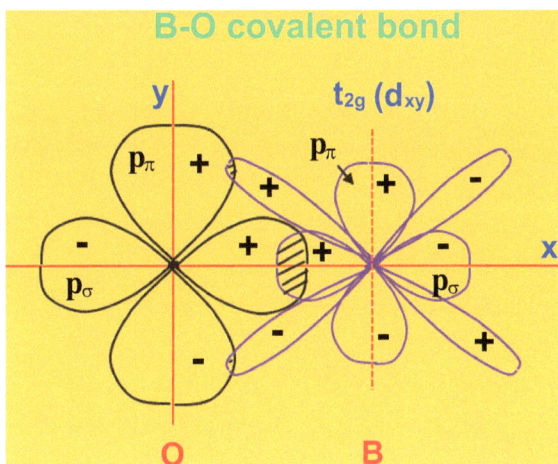

Fig. (4). The schematic of the B-O covalent bond formed in typical perovskite oxides. The electrons or holes are assumed to hop through these bonds while the oxidation state of B-site cations varies to contain the charges.

In order to extend ORR to the cathode surface, a certain ionic conductivity is essentially required. The intrinsic conduction of oxygen ions in oxide-ion conductors occurs *via* the oxygen-vacancy/oxygen-ion exchange mechanism. The movement of oxide ions in crystal structure from one lattice site to another one under the driving force of the electric field is a result of thermally activated hopping of oxygen ions. Three basic conditions must be satisfied: (a) adjacent vacancies provides the available positions for the dwelling of oxygen ions; (b) the oxygen ions are the largest specie with an ionic radius of 1.4 Å in the lattice. Therefore, the crystal structure should be very unique to allow the mobility of oxygen ions through the so-called "saddle point" where metal ions are distorted to vacate empty space; (c) the oxygen ions should be thermally activated to overcome the energy barrier for hopping from the sitting site to next vacant site. The ionic conductivity is dependent on the concentration of the oxygen vacancy induced by intrinsic defects and extrinsic defects caused by heat, impurity and also cation substitutions. In addition, the crystal structure of the materials also determines the mobility of oxygen ions from point of view of statistics.

Development of Low-temperature Cathode Materials

The Parameters for Characterizing the Excellence of Cathodes

In order to achieve lower SOFC operation temperature to $350 \sim 600 \,^{\circ}\text{C}$, a series of strategies have been adopted, such as using highly conductive electrolytes and active electrode materials, fabricating thin electrolyte film, and decorating electrode microstructure, *etc*. Among these strategies, developing new cathode materials is very promising to achieve low-temperature operation because the electrode polarization resistances are found to be dominant in total cell resistance in many cases while other fuel-cell components have been well developed to have less polarizations. Correspondingly, very extensive efforts have been made on the studying cathode materials, including different types of crystal structure with superior activity, element replacement to improve catalytic activity, conductivity, stability as well as compatibility, *etc*. Therefore, the proper characterizations of cathode are very important to evaluate the physical, chemical and electrochemical properties and to determine the feasibility of being a good candidate operated in low temperatures.

(a) Electrical Conductivity

A key parameter to judge a cathode is the electrical conductivity in air because electron or oxide-ion conduction is directly involved in the ORR on the cathode surface, particularly for MIEC. Generally, for many cases, total conductivities are measured by four-probe measurement technique, which can eliminate the resistance from wires and significantly decrease the measurement error. With higher conductivities, the cathode layer can be fabricated as thick as possible to increase reaction sites. Ionic conductivity is another useful parameter to see the nature of MIEC since it is only small fraction of the total. Technically, the ionic conductivity can be separated from total conductivity, such as concentration gradient cells, *etc.*

(b) Polarization Resistance

The electrochemical impedance spectroscopy (EIS) is a versatile technique to analyze the polarization contributions from different fuel-cell parts. The basic principal is that the sub-steps of ORRs are responsive to different frequency ranges when the small voltage fluctuation is applied. In full fuel-cell mode, the cathode polarization is impossibly separated from the low-frequency arc. Therefore, by studying the symmetric cells with configuration of electrolyte-support with cathode on both sides, the impedance response from such a cell can be regarded as the sole contribution from cathode material. It should be noted the resistance should be divided by two to obtain the correct values. Furthermore, by fitting the temperature-dependence nonlinear curves, the activation energy can be deduced by the Arrhenius equation, which is used for standard comparison with other cathode materials.

(c) Oxygen Surface Exchange and Diffusion Coefficient

For MIEC, the oxygen is absorbed on the cathode surface and then incorporate into the electrolyte through two pathways: the surface and bulk. The availability of bulk pathway results from the oxygen conduction of cathode. Therefore, in these ORR processes, two important material properties are involved: the incorporation of oxygen ions into electrolyte and the transport of oxygen ions within the bulk, correspond to the oxygen surface exchange coefficient (k) and the oxygen diffusion coefficient (D), respectively. There are two popular methods currently for measuring these two parameters, isotope exchange depth profiling (IEDP) and electrical conductivity relaxation (ECR). Coupled with the use of secondary ion mass spectrometry (SIMS) as the depth profiling technique, they can be experimentally derived. By fitting the normalized surface concentration profile of tracer element with solution to the diffusion equation, the D and k can be obtained when the experimental data accords with mathematical fitting plot. For the ECR technique, it is based on the theory that oxygen begins to diffuse into or out of the cathode bulk, causing an instant change of oxygen vacancy concentration when an abrupt change in oxygen partial pressure is applied [12].

By monitoring the instant change in electrical conductivity which is related with the change in oxygen vacancy concentration, the relaxation profiling plots can be mathematically analyzed and fitted to obtain the D and k values. For this method, the sample preparation, test conditions and also the mathematical model for fitting the experimental data can affect the accuracy.

High-Performance Cathodes Operating at Low Temperatures

There are various cathode materials developed for SOFCs and several of them are very active at low-temperature range (350~600 °C). Some review papers can be found to understand the development status of cathode materials. Herein, we emphasize on the cathode materials for LT-SOFCs.

(a) ABO$_3$-type Perovskite

The perovskite structured cathodes show very promising performance in oxide-ion or proton conducting SOFCs. Typically, they are developed based on $LaCoO_3$, $BaCoO_3$ and $SmCoO_3$ perovskite oxides. Zhan *et al.* reported an oxide hybrid, featuring a nanoporous $Sm_{0.5}Sr_{0.5}CoO_{3-\delta}$ (SSC) catalyst coated onto the internal surface of a high-porosity $La_{0.9}Sr_{0.1}Ga_{0.8}Mg_{0.2}O_{3-\delta}$ (LSGM) backbone by ion-infiltration method, exhibiting the very high catalytic activity for ORRs and thereby yielding low interfacial polarization resistance in air, *e.g.*, 0.021 ohm cm^2 at 650 °C and 0.043 ohm cm^2 at 600 °C [13]. The single fuel cell with SSC cathode exhibited high power densities of 2.02 W cm^{-2} at 650 °C and 1.46 W cm^{-2} at 600 °C when it is operated in humidified hydrogen and air. At lower temperatures, the cell can even show power densities of 0.91 W cm^{-2} and 0.47 W cm^{-2} at 550 °C and 500 °C, respectively.

Based on SSC cathode, the high performance at low temperatures has also been achieved by decorating cathode microstructure by infiltration technique, based on the $Sm_{0.2}Ce_{0.8}O_{1.9}$ (SDC) oxide-ion conducting electrolyte. Zhao and Chen *et al.* fabricated a nano-network of SSC cathode for low-temperature operation [14]. The nano-network consists of well-connected SSC nanowires which work as straight conducting pathways for conduction of oxygen-ion and electron, while the high porosity is well kept. In the preparation process, it was found that the formation of this unique nano-network structure was related to ramping rate in heating process. With a rate of 5 °C min^{-1}, SSC nanobeads with an average diameter of 55 nm were randomly distributed on the inner surface of the porous SDC electrolyte backbones. At 10 °C min^{-1}, the SSC nanobeads oppositely was decreased to 46 nm in diameter. Interestingly, the SSC nanowires were formed with composition of 5-8 nanobeads. The quicker ramping rate of 20 °C min^{-1}

yielded the similar structure. The performance of the fuel cell with SSC cathode prepared with a heating rate of 10 °C min^{-1} based on SDC electrolyte reached 0.44 and 0.81 W cm^{-2} at 500 and 600 °C, respectively. The interfacial electrode polarization resistance was reduced to as low as 0.21 ohm cm^2 at 500 °C, even lower than a BSCF-SDC composite cathode [15].

The SSC cathode was also used in proton conducting SOFCs. L. Yang and M. Liu *et al.* [16] reported a SSC and BZCY (BaZr$_{0.1}$Ce$_{0.7}$Y$_{0.2}$O$_{3-\delta}$) composite cathode with a number of active sites that facilitate the ORR involving H$^+$, O^{2-} and e' or h. on the active cathode surface area. The peak power densities reached 725, 598, 445 and 272 mW cm^{-2} at 700, 650, 600 and 500 °C, respectively.

In the last decade, cubic perovskite Ba$_{0.5}$Sr$_{0.5}$Co$_{0.8}$Fe$_{0.2}$O$_{3-\delta}$ (BSCF) oxide has gained increasing attention as a novel cathode material for intermediate-temperature (IT) SOFC applications [5, 17]. The A-site cation of BSCF perovskite is a typical alkaline-earth element rather than a rare-earth metal. This material was directly derived from SrCoO$_{3-\delta}$ perovskite which is an important parent compound for the development of many functional materials, particularly oxygen permeation membrane as might be interested for SOFC researches. For SrCoO$_{3-\delta}$ with cubic phase, it shows high electrical conductivity of ~160 S cm^{-1} at about 950 °C [18]. The reported ionic conductivity derived from the permeation flux data was as high as 2.5 S cm^{-1} at 900 °C, which is very encouraging.

Shao and Haile *et al.* firstly applied this BSCF oxide as a high-performance cathode with high power densities of 1010 and 402 mW cm^{-2} at 600 and 500 °C in hydrogen [5]. By measuring the polarization resistance from structural configuration of two-electrode symmetric cells with structural configuration of BSCF | electrolyte | BSCF, the ASR values describing all resistance terms associated with the electrode were collected in air. The determined ASR was very low: 0.055 ohm cm^2 at 600 °C and 0.51 ohm cm^2 at 500 °C, which can be further decreased when some current was applied. After this work reported, there have been extensive efforts on further decreasing the operating temperatures by different strategies [19 - 21].

(b) A$_2$B$_2$O$_{5+\delta}$-type Double Perovskite

In recent years, several groups have been focusing on studying the crystal chemistry and magnetic properties of LnBaCo$_2$O$_{5+\delta}$ (Ln=Pr, La, Nd, Sm, Gd, and Y) layered perovskite oxides [22 - 26]. The anisotropy resulting from the presence of cation ordered crystal structure has been reported to be able to improve the oxygen diffusion in comparison with non-ordered oxides. This class of materials is a very suitable as cathode candidates for LT-SOFCs due to the rapid oxygen surface exchange and lower activation energy for oxygen diffusion. These oxides are related to the "112"-type structure as firstly reported for $YBaFeCuO_5$ and consist of double layers of square pyramidally coordinated cations of cobalt. The unit cell of $LnBaCo_2O_{5+\delta}$ has been reported to be orthorhombic structure with dimensions $a_p \times 2a_p \times 2a_p$ or $a_p \times a_p \times 2a_p$ (space group P_{mmm}), or tetragonal structure with dimensions $a_p \times a_p \times 2a_p$ ($P_{4/mmm}$), where a_p refers to the lattice parameter of the primitive cubic unit cell. The symmetry and unit cell size highly depend on the oxygen content, the ionic size of Ln and temperature [27].

The first structural feature of layered perovskite is the ordering of lanthanide and Ba ions in alternating (001) layers [28]. The doubling of a_p along the c-axis is due to the location of Ln and Ba cations in alternative layers perpendicular to the c-axis; while the doubling of a_p along the b-axis results from the periodic locations of the oxygen vacancies along the a-direction. Another feature of layered perovskite is the ordered distribution of oxygen vacancies in the Ln layers. In order to tailor the transport properties, the oxygen non-stoichiometry can be changed by means of heat treatment. When the oxygen content is $5+\delta \approx 5.5$, the orthorhombic symmetry appears. In this case, a perfectly ordered sample would present 100% of the oxygen occupancy at Wyckoff position 1g, located at (0, 1/2, 1/2), and 0% at the 1c position of (0, 0, 1/2). The results with neutron powder diffraction (NPD) technique reported by Frontera *et al.* [29] show that the real oxygen occupancies are 91% and 12%, respectively. On the other hand, when the oxygen content is very different from 5.5, the crystal structure becomes tetragonal with the LnO layers partially empty.

The oxygen self-diffusion and surface reactions correlating with ORR of cathode materials can be significantly enhanced because of reduced oxygen bond strength leading to disorder-free pathways for fast oxygen ion diffusion, which is responsible for the higher surface oxygen exchange rate (k*) observed in the ordered $GdBaMn_2O_{5+\delta}$, compared to the disordered $Gd_{0.5}Ba_{0.5}MnO_{3-x}$ with single perovskite phase [27]. The electronic conduction in $LnBaCo_2O_{5+\delta}$ occurs *via* hopping mechanism along the Co-O-Co bonds through a double exchange process similar to other simple perovskite ABO_3-type cathodes. Kilner *et al.* have investigated $GdBaCo_2O_{5+x}$ as a SOFC cathode [30]. With a 15-µm thick electrolyte of YSZ, the anode-supported fuel cell showed maximum power density of 300 mW cm^{-2} at 700 °C in H_2. The cell performance has been further improved by adopting doped ceria as the electrolyte and composite cathode [31]. Kim *et al.* have found that $PrBaCo_2O_{5+\delta}$ (PBCO) oxide possesses rapid oxygen transport kinetics at low temperatures (300-500 °C) measured by electrical conductivity relaxation (ECR) and isotope exchange as well as depth profiling (IEDP) on bulk ceramics and highly epitaxial PBCO thin films prepared by pulsed laser deposition (PLD) technique [32]. In order to obtain low electrode polarization resistances, it is required that the value of k*D* is greater than 10^{-14} cm^3s^{-2} and k*/D* smaller than 100 cm^{-1} [33, 34].

The electrochemical performance of PBCO cathode was evaluated based on oxide-ion conducting electrolyte of SDC [35]. The maximum power densities of 866, 583, 313 and 115 mW cm^{-2} were achieved at 650, 600, 550 and 500 °C, respectively, when hydrogen is used as fuel and ambient air as oxidant. PBCO also showed excellent electrochemical performance in proton-conducting SOFCs. Liu *et al.* [36] reported a peak power density of 545 mW cm^{-2} at 700 °C in H_2 due to the low electrode polarization resistance of 0.15 ohm cm^2. Recently, Ding *et al.* reported a nanospike structured cathode prepared by a novel low-temperature sintering process followed with a discharge treatment [37, 38]. The voltage is illustrated to enable the growth of interfacial nanospikes on the surface of porous cathode bulk. The fuel cell exhibiting superior electrochemical performance allows to be operated at low operation temperatures (350-550 °C) and also meets the requirements for long-term stability. The cell exhibited maximum power densities (P_{max}) of 1.453, 1.044, 0.657 and 0.316 W cm^{-2} at 550, 500, 450, and 400 °C, respectively. The P_{max} of 0.135 W cm^{-2} could be achieved at the temperature of 350 °C, the lowest testing temperature of SOFCs in open literature.

CONCLUDING REMARKS

The technical requirements, working mechanism and a series of oxide materials for cathodes operating at low-temperature range (350~600 °C) have been discussed in this chapter. The criteria of material selection and understanding what affects the intrinsic properties for ORR are fundamental theory preparation for anyone who are interested in developing advanced functional cathodes for low-temperature SOFCs. Perovskite-type oxides are typical class catalysts of fast ORR while nanostructure-engineering is a very useful route to increase the active surface for the complex multi-steps of reaction. The high tolerance factor against structure distortion allows the crystal to be tuned by cation substitution in order to create the essential defects for surface chemistry reactions. Furthermore, the special crystal structure yielding fast oxygen surface exchange and bulk diffusion of layered perovskites is demonstrated to facilitate ORRs through fast channel of oxygen diffusion while simple ABO_3 perovskite is transformed into double perovskite in which the A-site cations show sequential stacking and oxygen vacancies are orderly distributed on the special planes of lanthanide ions. It should be particularly noted that the low-temperature SOFCs can not only be achieved by cathode, but also optimization of electrolyte material, anode microstructure and electrode/electrolyte interfaces. The mathematical modeling is a versatile tool to explore the trends and anticipate the potential factors that is correlated with the physical, kinetics or electrochemical phenomena occurring in SOFCs, where the mass and heat transports are involved, although this is not mentioned. Considering the long-term operation of such SOFC device, more factors should be taken into account to evaluate cathodes, such as thermal, chemical stability and compatibility, *etc*. While this chapter only focuses on some specific aspects, other useful literature about cathode review can be also found.

CONFLICT OF INTEREST

The author confirms that author has no conflict of interest to declare for this publication.

ACKNOWLEDGEMENTS

Declared none.

REFERENCES

[1]   Richter, J.; Holtappels, P.; Graule, T.; Nakamura, T.; Gauckler, L.J. Materials design for perovskite SOFC cathodes. *Monatsh. Chem.,* **2009**, *140*, 985-999.
       [http://dx.doi.org/10.1007/s00706-009-0153-3]

[2]   Mogensen, M.; Lybye, D.; Bonanos, N.; Hendriksen, P.V.; Poulsen, F.W. Factors controlling the oxide ion conductivity of fluorite and perovskite structured oxides. *Solid State Ion.,* **2004**, *174*, 279-286.
       [http://dx.doi.org/10.1016/j.ssi.2004.07.036]

[3]   Murray, E.P.; Barnett, S.A. (La,Sr)MnO$_3$-(Ce,Gd)O$_{2-x}$ composite cathodes for solid oxide fuel cells. *Solid State Ion.,* **2001**, *143*, 265-273.
       [http://dx.doi.org/10.1016/S0167-2738(01)00871-2]

[4]   Uchida, H.; Arisaka, S.; Watanabe, M. High performance electrode for medium-temperature solid oxide fuel cells La(Sr)CoO$_3$ cathode with ceria interlayer on zirconia electrolyte. *Electrochem. Solid-State Lett.,* **1999**, *2*, 428-430.
       [http://dx.doi.org/10.1149/1.1390860]

[5]   Shao, Z.; Haile, S.M. A high-performance cathode for the next generation of solid-oxide fuel cells. *Nature,* **2004**, *431*(7005), 170-173.
       [http://dx.doi.org/10.1038/nature02863] [PMID: 15356627]

[6]   Fleig, J. Solid oxide fuel cell cathodes: Polarization mechanisms and modeling of the electrochemical performance. *Annu. Rev. Mater. Res.,* **2003**, *33*, 361-382.
       [http://dx.doi.org/10.1146/annurev.matsci.33.022802.093258]

[7]   Norskov, J.K.; Rossmeisl, J.; Logadottir, A.; Lindqvist, L. Origin of the overpotential for oxygen reduction at a fuel-cell cathode. *J. Phys. Chem. B,* **2004**, *108*, 17886-17892.
       [http://dx.doi.org/10.1021/jp047349j]

[8] Stamenkovic, V.R.; Fowler, B.; Mun, B.S.; Wang, G.; Ross, P.N.; Lucas, C.A.; Marković, N.M. Improved oxygen reduction activity on $Pt_3Ni(111)$ *via* increased surface site availability. *Science,* **2007**, *315*(5811), 493-497.
[http://dx.doi.org/10.1126/science.1135941] [PMID: 17218494]

[9] Stamenkovic, V.; Mun, B.S.; Mayrhofer, K.J.; Ross, P.N.; Markovic, N.M.; Rossmeisl, J.; Greeley, J.; Nørskov, J.K. Changing the activity of electrocatalysts for oxygen reduction by tuning the surface electronic structure. *Angew. Chem. Int. Ed. Engl.,* **2006**, *45*(18), 2897-2901.
[http://dx.doi.org/10.1002/anie.200504386] [PMID: 16596688]

[10] Suntivich, J.; Gasteiger, H.A.; Yabuuchi, N.; Nakanishi, H.; Goodenough, J.B.; Shao-Horn, Y. Design principles for oxygen-reduction activity on perovskite oxide catalysts for fuel cells and metal-air batteries. *Nat. Chem.,* **2011**, *3*(7), 546-550.
[http://dx.doi.org/10.1038/nchem.1069] [PMID: 21697876]

[11] Khan, W.; Naqvi, A.H.; Gupta, M.; Husain, S.; Kumar, R. Small polaron hopping conduction mechanism in Fe doped $LaMnO_3$. *J. Chem. Phys.,* **2011**, *135*(5), 054501.
[http://dx.doi.org/10.1063/1.3615720] [PMID: 21823706]

[12] Elshof, J.E.; Lankorst, M.H.; Bouwmeester, H.J. Oxygen exchange and diffusion coefficients of strontium-doped lanthanum ferrites by electrical conductivity relaxation. *J. Electrochem. Soc.,* **1997**, *144*, 1060-1067.
[http://dx.doi.org/10.1149/1.1837531]

[13] Da Han, ; Liu, X.; Zeng, F.; Qian, J.; Wu, T.; Zhan, Z. A micro-nano porous oxide hybrid for efficient oxygen reduction in reduced-temperature solid oxide fuel cells. *Sci. Rep.,* **2012**, *2*, 462.
[http://dx.doi.org/10.1038/srep00462] [PMID: 22708057]

[14] Zhao, F.; Wang, Z.Y.; Liu, M.F.; Zhang, L.; Xia, C.R.; Chen, F.L. Novel nano-network cathodes for solid oxide fuel cells. *J. Power Sources,* **2008**, *185*, 13-18.
[http://dx.doi.org/10.1016/j.jpowsour.2008.07.022]

[15] Ai, N.; Lu, Z.; Chen, K.F.; Huang, X.Q.; Wei, B.; Zhang, Y.H.; Li, S.Y.; Xin, X.S.; Sha, A.Q.; Su, W.H. Low temperature solid oxide fuel cells based on $Sm_{0.2}Ce_{0.8}O_{1.9}$ films fabricated by slurry spin coating. *J. Power Sources,* **2006**, *159*, 637-640.
[http://dx.doi.org/10.1016/j.jpowsour.2005.11.057]

[16] Yang, L.; Zuo, C.D.; Wang, S.Z.; Cheng, Z.; Liu, M.L. A novel composite cathode for low-temperature SOFCs based on oxide proton conductors. *Adv. Mater.,* **2008**, *20*, 3280-3283.
[http://dx.doi.org/10.1002/adma.200702762]

[17] Zhou, W.; Ran, R.; Shao, Z.P. Progress in understanding and development of $Ba_{0.5}Sr_{0.5}Co_{0.8}Fe_{0.2}O_{3-\delta}$-based cathodes for intermediate-temperature solid-oxide fuel cells: A review. *J. Power Sources,* **2009**, *192*, 231-246.
[http://dx.doi.org/10.1016/j.jpowsour.2009.02.069]

[18] Deng, Z.Q.; Wang, W.S.; Liu, W.; Chen, C.S. Relationship between transport properties and phase transformations in mixed-conducting oxides. *J. Solid State Chem.,* **2006**, *179*, 362-369.
[http://dx.doi.org/10.1016/j.jssc.2005.10.027]

[19] Zeng, P.Y.; Chen, Z.H.; Zhou, W.; Gu, H.X.; Shao, Z.P.; Liu, S.M. Re-evaluation of $Ba_{0.5}Sr_{0.5}Co_{0.8}Fe_{0.2}O_{3-\delta}$ perovskite as oxygen semi-permeable membrane. *J. Membr. Sci.,* **2007**, *291*, 148-156.
[http://dx.doi.org/10.1016/j.memsci.2007.01.003]

[20] Wei, B.; Lü, Z.; Huang, X.Q.; Miao, J.P.; Sha, X.Q.; Xin, X.S.; Su, W.H. Crystal structure, thermal expansion and electrical conductivity of perovskite oxides $Ba_xSr_{1-x}Co_{0.8}Fe_{0.2}O_{3-\delta}$ ($0.3 \leq x \leq 0.7$). *J. Eur. Ceram. Soc.,* **2006**, *26*, 2827-2832.
[http://dx.doi.org/10.1016/j.jeurceramsoc.2005.06.047]

[21]   Liu, Q.L.; Khor, K.A.; Chan, S.H. High-performance low-temperature solid oxide fuel cell with novel BSCF cathode. *J. Power Sources,* **2006**, *161*, 123-128.
[http://dx.doi.org/10.1016/j.jpowsour.2006.03.095]

[22]   Maignan, A.; Martin, C.; Pelloquin, D.; Nguyen, N.; Raveau, B. Structural and magnetic studies of ordered oxygen-deficient perovskites *Ln* $BaCo_2O_{5+\delta}$, closely related to the '112' structure. *J. Solid State Chem.,* **1999**, *142*, 247-260.
[http://dx.doi.org/10.1006/jssc.1998.7934]

[23]   Martin, C.; Maignan, A.; Pelloquin, D.; Nguyen, N.; Raveau, B. Magnetoresistance in the oxygen deficient $LnBaCo_2O_{5.4}$ (Ln=Eu, Gd) phases. *Appl. Phys. Lett.,* **1997**, *71*, 1421-1423.
[http://dx.doi.org/10.1063/1.119912]

[24]   Moritomo, Y.; Takeo, M.; Liu, X.J.; Akimoto, T.; Nakamura, A. Metal-insulator transition due to charge ordering in $R_{1/2}Ba_{1/2}CoO_3$. *Phys. Rev. B,* **1998**, *58*, R13334-R13337.
[http://dx.doi.org/10.1103/PhysRevB.58.R13334]

[25]   Vogt, T.; Woodward, P.M.; Karen, P.; Hunter, B.A.; Henning, P.; Moodenbaugh, A.R. Low to high spin-state transition induced by charge ordering in antiferromagnetic $YBaCo_2O_5$. *Phys. Rev. Lett.,* **2000**, *84*(13), 2969-2972.
[http://dx.doi.org/10.1103/PhysRevLett.84.2969] [PMID: 11018988]

[26]   Burley, J.C.; Mitchell, J.F.; Short, S.; Miller, D.; Tang, Y. Structural and magnetic chemistry of $NdBaCo_2O_{5+\delta}$. *J. Solid State Chem.,* **2003**, *170*, 339-350.
[http://dx.doi.org/10.1016/S0022-4596(02)00101-9]

[27]   Taskin, A.; Lavrov, A.; Ando, Y. Transport and magnetic properties of $GdBaCo_2O_{5+x}$ single crystals: A cobalt oxide with square-lattice CoO_2 planes over a wide range of electron and hole doping. *Phys. Rev. B,* **2005**, *71*, 134414-134441.
[http://dx.doi.org/10.1103/PhysRevB.71.134414]

[28]   Moritomo, Y.; Akimoto, T.; Takeo, M.; Machida, A.; Nishibori, E.; Takata, M.; Sakata, M.; Ohoyama, K.; Nakamura, A. Metal-insulator transition induced by a spin-state transition in $TbBaCo_2O_{5+\delta}$ (δ=0.5). *Phys. Rev. B,* **2000**, *61*, R13325-R13328.
[http://dx.doi.org/10.1103/PhysRevB.61.R13325]

[29]   Frontera, C.; Caneiro, A.; Carrillo, A.E.; Oró-Solé, J. García-Mun~oz JL. Tailoring oxygen content on $PrBaCo_2O_{5+\delta}$ layered cobaltites. *Chem. Mater.,* **2005**, *17*, 5439-5445.
[http://dx.doi.org/10.1021/cm051148q]

[30]   Tarancón, A.; Morata, A.; Dezanneau, G.; Skinner, S.J.; Kilner, J.A.; Estradé, S. H.-Ramírez F, Peiró R, Morante JR. $GdBaCo_2O_{5+x}$ layered perovskite as an intermediate temperature solid oxide fuel cell cathode. *J. Power Sources,* **2007**, *174*, 255-263.
[http://dx.doi.org/10.1016/j.jpowsour.2007.08.077]

[31]   Lee, Y.; Kim, D.Y.; Choi, G.M. $GdBaCo_2O_{5+x}$ cathode for anode-supported ceria SOFCs. *Solid State Ion.,* **2011**, *192*, 527-530.
[http://dx.doi.org/10.1016/j.ssi.2010.04.027]

[32]   Kim, G.; Wang, S.; Jacobson, A.J.; Yuan, Z.; Donner, W.; Chen, C.L.; Reimus, L.; Brodersen, P.; Mins, C.A. Oxygen exchange kinetics of epitaxial $PrBaCo_2O_{5+\delta}$ thin films. *Appl. Phys. Lett.,* **2006**, *88*, 024103.
[http://dx.doi.org/10.1063/1.2163257]

[33]   Tarancón, A.; Skinner, S.J.; Chater, R.J. H.-Ramírez F, Kilner JA. Layered perovskites as promising cathodes for intermediate temperature solid oxide fuel cells. *J. Mater. Chem.,* **2007**, *17*, 3175-3181.
[http://dx.doi.org/10.1039/b704320a]

[34]   Tarancón, A.; Burriel, M.; Santiso, J.; Skinner, S.J.; Kilner, J.A. Advanced in layered oxide cathodes for intermediate temperature solid oxide fuel cells. *J. Mater. Chem.,* **2010**, *20*, 3799-3813.
[http://dx.doi.org/10.1039/b922430k]

[35]   Zhu, C.J.; Liu, X.M.; Yi, C.S.; Yan, D.T.; Su, W.H. Electrochemical performance of $PrBaCo_2O_{5+\delta}$ layered perovskite as an intermediate-temperature solid oxide fuel cell cathode. *J. Power Sources,* **2008**, *185*, 193-196.
[http://dx.doi.org/10.1016/j.jpowsour.2008.06.075]

[36]   Zhao, L.; He, B.B.; Lin, B.; Ding, H.P.; Wang, S.L.; Ling, Y.H.; Peng, R.R.; Meng, G.Y.; Liu, X.Q. High performance of proton-conducting solid oxide fuel cell with a layered $PrBaCo_2O_{5+\delta}$ cathode. *J. Power Sources,* **2009**, *194*, 835-837.
[http://dx.doi.org/10.1016/j.jpowsour.2009.06.010]

[37]   Ding, H.P.; Xue, X.J. An interfacial nanospike-structured cathode for low temperature solid oxide fuel cells. *Adv. Mater. Interfaces,* **2015**, *1*, 1400008.
[http://dx.doi.org/10.1002/admi.201400008]

[38]   Ding, H.P. *Material synthesis and fabrication method development for intermediate temperature solid oxide fuel cells.,* Doctoral dissertation, http://scholarcommons.sc.edu/etd/2622, **2014**.

A Brief Review on Anode Materials and Reactions Mechanism in Solid Oxide Fuel Cells

Caroline Gomes Moura[1,*], João Paulo de Freitas Grilo[2], Rubens Maribondo do Nascimento[3,*] and Daniel Araújo de Macedo[4]

[1] *Department of Mechanical Engineering, University of Minho, Braga, Portugal*

[2] *Department of Materials and Ceramic Engineering/CICECO, University of Aveiro, Aveiro, Portugal*

[3] *Department of Materials Engineering, Federal University of Rio Grande do Norte, Natal, Brazil*

[4] *Department of Materials Engineering, Federal University of Paraíba, João Pessoa, Brazil*

Abstract: This chapter presents a state-of-art brief review on anode materials for SOFC. Materials, processing and synthesis routes to attain porous anodes are highlighted. Especial attention is given to Ni-ceramic phase (especially fluorite-type structure ceramics) cermets. Some aspects about prospects and problems of the currently developed electrodes materials are elucidated. Electrodes for intermediate temperature SOFCs (IT-SOFCs) are discussed in relation to other conventional electrodes. The electrochemical characterization of anodes, as mixed ionic-electronic conductors, is briefly outlined.

Keywords: Anodes, Ceramics, Cermets, chemical routes, conventional routes, Electrical properties, Electrodes, Mixed Ionic-Electronic Conductor, Porous, Solid Oxide Fuel Cell.

INTRODUCTION

The anode is the electrode where the fuel oxidation occurs. As the cathode, this component must also exhibit high electronic conductivity, good catalytic activity for the fuel oxidation reactions and sufficient porosity to allow the transport of fuel to the anode/electrolyte interface and the removal of reaction products. In addition, the anode should be chemically stable and thermally compatible with the other SOFC components [1, 2].

* **Corresponding author Caroline G. Moura and Rubens M. Nascimento:** Department of Mechanical Engineerin/CMEMS, University of Minho, Guimarães, Portugal; Tel/Fax: +351 253 601 100; E-mail: caroline.materiais@gmail.com, Department of Materials Engineering, Federal University of Rio Grande do Norte, Natal, Brazil; Tel/Fax: +55 84 3215-3883; E-mail: rubens.maribondo@gmail.com

Moisés R. Cesário & Daniel A. de Macedo (Eds.)
All rights reserved-© 2017 Bentham Science Publishers

The electrochemical performance of the anode depends on the charge transport resistance (electrons and ions), inside the anode and the anode/electrolyte interface, and the resistance of gas transport. The increase of the triple phase boundaries (TPB) length, by microstructural optimization and phase composition, are the most efficient ways to improve the electrochemical performance of anodes [3, 4].

Internal reform and tolerance to sulfur-containing compounds are also essential to the anodes, especially when a hydrocarbon fuel is used, *e.g.* methane. The porosity of the anodes is a very important factor, not only because it is related to high densities of triple phase boundaries, but also because it avoids mass transport limitation. In this regard, many studies have reported the use of pore formers (graphite, starch, citric acid, *etc.*) in order to obtain suitable porosity in anodes [5, 6]. However, due to the tendency to agglomerate of pore-forming agents, it is sometimes difficult to ensure good structural performance and permeation of gases in these electrodes [7].

Anode Materials

Porous metal electrodes are the most suitable materials to work in reducing atmosphere conditions. These materials have good electronic conductivity and good fuel permeability. Metals such as Ag, Pt, Ru, Co, Fe, and Mn have been tested as anode materials [8 - 11]. In these studies it was observed that platinum lost adhesion with the electrolyte after a few hours of cell operation. Nickel, despite its low cost, good chemical stability and excellent catalytic activity for oxidation of hydrogen, had a serious drawback with microstructural stability. Ni has very low melting point and sintering temperature (1453 and 1000 °C respectively) which enable the grain growth during operation, this behavior can difficulty the passage of gases. Ni also has poor adhesion on dense electrolytes. Furthermore, pure Ni has an incompatible thermal expansion coefficient (TEC) with that of the most used electrolyte materials, YSZ and GDC (gadolinium doped ceria) (13.3 x 10^{-6} K^{-1}, 10.5 x 10^{-6} K^{-1}, 12.0 x 10^{-6} K^{-1} respectively) [12, 13].

Spacil proposed an easy way to solve the grain growth issue in Ni based anodes [14]. In his patent, Spacil proposed a mixture of YSZ and nickel particles (Ni), creating the Ni – YSZ cermet. This new material had superior properties than those of pure nickel electrodes. Nickel, characterized by its high electrocatalytic activity and relatively low price, makes the anode an electronic conductor with the possibility to perform steam reforming operations [15]. The ceramic phase basically suppress grain growth during cell operation, which prevents electrode pores obstruction; and improve the thermal and chemical compatibility at the electrolyte/electrode interface. In addition, the presence of the ceramic phase creates TPB in the entire anode extension where the fuel oxidation reaction occurs [16, 17]. Ni – YSZ cermets became the reference for SOFC anodes.

The idea of using cermets as anode materials began to be adopted in the development of new SOFC systems containing electrolytes with properties higher than YSZ. Examples of these materials are doped ceria, lanthanum gallates, and barium zirconate/barium cerate. For each SOFC system based on a new electrolyte, nickel can be mix with the corresponding electrolyte material, forming a Ni cermet – electrolyte. In these two-phase systems, the percolation limit is about 30 vol % of the metal phase (Ni), it is a required content to occur the transition from predominantly ionic to predominantly electronic conductivity. Furthermore, these materials are used in order to minimize the operating temperature of SOFCs, restricting the possibility of grain growth during the operation of the cell [18].

In recent years, the replacement of YSZ for doped ceria in nickel based cermets has been an increasingly common practice in the development of SOFC anodes. Compared to Ni-YSZ, Ni-GDC cermets have several advantages especially because of the properties of ceria based compounds (GDC has ionic conductivity at 800 °C equivalent to that of YSZ at 1000 °C, approximately 10^{-1} S/cm) which become mixed conductors in an atmosphere of hydrogen [19, 20]. As a mixed conductor, ceria can reduce the ions and transport electrons, promoting the charge transfer reaction in the entire area of electrode/gas interface. A notable feature of ceria-based anodes in relation to those containing YSZ is the ability to resist to carbon deposition on Ni surface, mainly when using methane-rich atmospheres [21, 22]. The good thermal compatibility with the ceria-based electrolytes is another great advantage of Ni-doped ceria cermets.

The electrochemical activity of these materials is strongly influenced by the electrical conductivity of each component and microstructural parameters such as grain size, composition, phase distribution, and connectivity between the ionic conductor (doped ceria), catalytic metal (Ni) and pore. The microstructure is influenced, among other factors, by the characteristics of the starting powders. The use of nanoparticles provides significant microstructural changes, especially increasing the number of contacts between Ni and the ceramic phase, which implies in extension of the TPB area and, as a result, improvement of the electrochemical performance. Considerable electrode microstructural changes have been observed by varying the starting powders [23 - 25]. The improvement of the electrochemical performance by increasing the TPB length can be provided by manufacturing nanostructured anode materials. This type of material has attracted great interest due to their significantly superior properties compared to the most common materials. If Ni and CGO grain sizes decrease on the anode microstructure, the contact between these phases increases, thereby increasing the contact with the gas phase, which improves the performance of the anode [26 - 30].

NiO-GDC Composites

The preparation of advanced ceramic/cermet components with high added value and specific functions, as SOFCs electrodes, requires the use of powders with strict control of the chemical composition and high homogeneity in the distribution of dopants in the crystal lattice. For example, Chavan *et al.* [31] reported the influence of the NiO content on the activation energy, as shown in Fig. (**1**). As can be seen, the activation energy decreases as NiO content increases. This phenomenon is only observed in NiO-GDC microstructures exhibiting high homogeneity and suitable phase distribution. The inserts in Fig. (**1**) show examples of NiO-GDC microstructures with NiO content below (A) and above (B) the percolation limit. The lighter grains are GDC, and the darker grains are NiO. With low NiO content, GDC grains are more significantly connected (insert A), which increases the activation energy due to the ionic nature of the conductivity. For higher NiO contents, above the percolation limit (40 mol % of NiO), the activation energy decreases, suggesting that NiO and GDC are homogeneously distributed. Hence, NiO tridimensional contacts are better established in the insert B, which can be associated with lower activation energies [31].

The fabrication of Ni – GDC cermets is prepared by mechanical mixing of NiO and GDC powders derived from chemical routes or even commercial powders. The resulting material of the mixture that could be performed by ball milling and/or high energy milling, is shaped, sintered in oxidizing atmosphere, and finally submitted to heat treatment in reducing atmosphere, becoming Ni-doped ceria cermets [31 - 36].

Despite its simplicity and good control of chemical composition, the conventional method of mixing oxides is greatly influenced by the characteristics of the starting powders and the processing route. This dependence in most cases involves non-uniform distribution of components and consequently non-homogeneous microstructure. Aiming to minimize this problem, many researchers have proposed alternative preparation routes, especially chemical methods, to obtain nanocomposite anode powders. The main purpose has been to synthesize nanocomposites with uniform distribution of phases [7, 31]. The use of nanocomposites, allowing a better distribution of Ni in the ceria matrix, has been appointed as one of the key approaches to reduce the Ni grain growth due to sintering and SOFC operating, enhancing electrode stability and performance [37, 38].

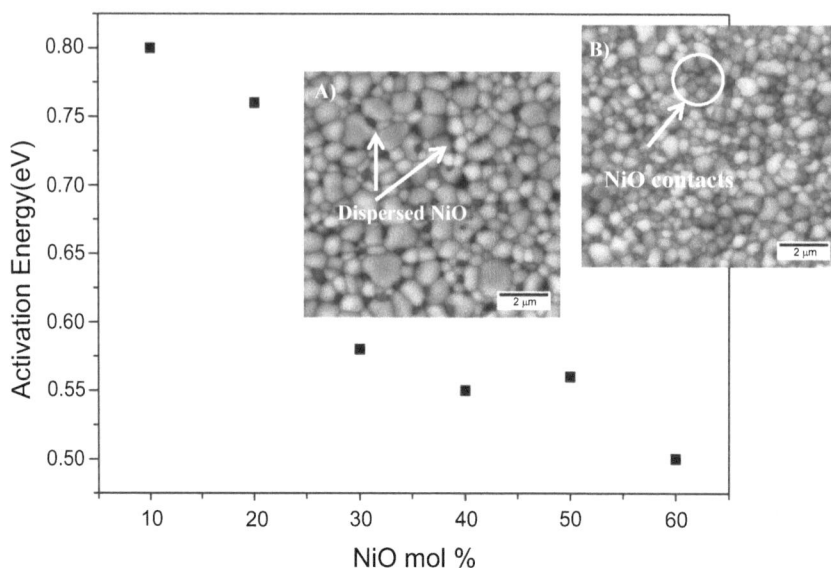

Fig. (1). NiO content influence on the activation energy. Values reported in [31].

Several chemical routes to obtain NiO – doped ceria nanopowders have been reported in literature. The effects of the preparation method on the powder morphology and phase distribution are the main factors to be considered before the choice of the synthesis method. Table **1** presents a brief overview of properties of Ni-GDC electrodes derived from powders synthesize by different methods.

Table 1. Brief overview of reported works using Ni-GDC electrodes derived from powders synthesize by different methods.

Reported Work	Powder Manufacturing Method	Ratio NiO – GDC	Brief Overview
[39]	Mixture of commercial powders	65-35 wt%	GDC and LSCF (Lanthanum strontium cobaltite ferrite) films were printed onto both sides of anode-supported single cell fabricated by tape casting method. The anode microstructure had suitable microstructure even after cell testing. The single cell maximum power density of 909, 623, 335 and 168 mW.cm^{-2} was obtained at 650, 600, 550 and 500 °C, respectively.

(Table 1) contd.....

Reported Work	Powder Manufacturing Method	Ratio NiO – GDC	Brief Overview
[40]	Polymeric organic complex solution method	50 – 50, 45–55 and 40–60 wt%.	NiO-GDC powders were pressed in pellets and sintered. High relative densities of the samples, especially 50 – 50 wt% composition, were achieved. Furthermore, after reducing in H_2 atmosphere, this composition showed good connection between Ni and GDC grain and interconnected porosity, which is essential to increase the TPB.
[41]	Cellulose-precursor method	-	NiO-GDC mixture with 80 nm particle size, had an uniform phase distribution. The electrochemical performance of the material, using lanthanum gallate as electrolytes, was later evaluated by polarization measurements in three electrodes.
[42]	electrostatic-assisted ultrasonic spray pyrolysis electrostatic-assisted ultrasonic spray pyrolysis Electrostatic-assisted ultrasonic spray pyrolysis	70 – 30, 60 – 40 and 50 – 50 mol%	NiO-GDC film was deposited in GDC substrate evaluating the effects of deposition temperature, electric field strength and composition. After reducing in H2 atmosphere, 5:5 composition porous film at deposition temperature of 450 ºC and 12 kV applied voltage yielded the most TPB junctions. Ni:GDC (6:4) yielded the lowest impedances, ASR at 550 ºC was 0.09 Ω. cm^2.
[25]	Hydroxide co-precipitation method	-	LSCF nanoparticles were screen printed onto GDC/NiO-GDC co-pressed and co-sintered sample in order to fabricate the single cell. Particle size of 10 – 40 nm was obtained for the NiO-GDC composite nanopowders. Its single cell exhibits an OCV of 0.935 V and the maximum power densities were 134, 272, 475 and 713 $mW.cm^{-2}$ at 500, 550, 600 and 650 ºC, respectively.
[43]	Hydroxide co-precipitation method and mixture of commercial powders	60 – 40 wt%	Single cell was fabricated by screen printing LSCF nanoparticles onto GDC/NiO-GDC co-pressed and co-sintered half-cell using two types of NiO-GDC powders. More uniform and homogeneous microstructure was obtained for the nanocomposites obtained by the chemical route. The maximum power density of 360 $mW.cm^{-2}$ for the cell with the anode prepared by chemical route, compared to 89 $mW.cm^{-2}$ for the cell with commercial anode, both measured at 600 ° C.
[32]	Glycine–nitrate process and mixture of commercial powders	65 – 35 wt%	Four types of GDC electrolyte supported cells were fabricated with the mixture of the powders prepared by chemical route and the commercial powders using $Sm_{0.5}Sr_{0.5}CoO_3$ + GDC as cathode. The lowest impedance was 0.06 $\Omega.cm^2$ for powders prepared by the synthesis. The OCVs and maximum power densities were 0.94 – 0.83 V and 220 – 402 $mW.cm^{-2}$ at 600 and 700 ºC respectively for the same type of anode.

(Table 1) contd.....

Reported Work	Powder Manufacturing Method	Ratio NiO – GDC	Brief Overview
[31]	Solution combustion	NiO$_x$-GDC$_{(1-x)}$ where x = 0.1, 0.2, 0.3, 0.4, 0.5, 0.6 (in mol %)	GDC and NiO powders obtained by chemical rout were mixed and pressed into pellets and sintered. Relative densities decrease with NiO content, as well as, the activation energy (between 500-700 ºC). Nano-composites with 30–60 mol % NiO showed a porous microstructure composed of uniformly distributed and well-connected constituent grains. Percolative behavior in NiO–GDC nanocomposite is exhibited in compositions as near as 40 mol % of NiO.
[44]	Hydroxide and oxalate reverse co-precipitation methods (comparative study)	60 – 40 wt %	NiO – GDC powders obtained by the two type of synthesis was pressed in pellets and sintered. Average particles size of less than 25 nm were obtained by both type of synthesis. NiO-GDC nanocomposite powders synthesized by the hydroxide reverse co-precipitation method have better sinterability than the NiO– GDC nanocomposite powders synthesized by the oxalate reverse co-precipitation method.
[45]	Polymeric precursor method (one and two steps – comparative study)	50 – 50 wt %	NiO-GDC powders were obtained by one step (precursor resins of the CGO and NiO phases obtained by the polymeric precursor method and mixed to become one homogeneous resin) and two step (single phase powders obtained by the same route were mixed mechanically) synthesis. Single cells YSZ electrolyte supported were fabricated printing NiO-GDC and LSFC-SDC onto both sides. One step synthesis achieved smaller average particle sizes, homogeneity and absence of agglomeration or phase segregation. The maximum power densities were ~ 30 – 52 mW.cm^{-2} for two and one step synthesis respectively, at 850 ºC.
[46]	Polymeric precursor method (one step) and mixture of commercial powders (comparative study)	50 – 50 wt%	NiO-GDC powders obtained by one step synthesis [45] and commercial powders mixture were printed onto both side of GDC electrolyte supported single cell. One-step anodes yielded better microstructure and electrochemical performance than conventional anodes prepared by commercial powders mixture. The OCVs and maximum current densities were 95- 91 mV and 215-322 mA.cm^{-2} for anodes obtained by conventional powder mixing and one step synthesis respectively.

As can be seen, Table **1** shows different results which intrinsically depends on the powder manufacturing method. A good combination of the synthesis method and processing route can provide high performance Ni-GDC cermet anodes. In many cases, NiO-GDC composite powders obtained by chemical routes have better properties than those fabricated by the conventional mechanical mixing method. For example, Zha *et al.* [32] reported that the lowest impedance for anodes prepared using powders derived from a glycine–nitrate process is 0.06 $\Omega.cm^2$ which is a remarkable result when compared to the impedance of electrode prepared from commercial powders, 1.61 $\Omega.cm^2$. These results can be related with good values of maximum power densities, as can be seen in [32]. Cela and Macedo [45, 46] also reported improved results for powders obtained by a one step synthesis approach. The maximum power density for a single cell fabricated with one step synthesis NiO-GDC composite is almost twice higher than that fabricated with a two steps approach (~ 52 – 30 mW.cm^{-2} respectively) [45].

ELECTROCHEMICAL PROCESSES IN SOFC ELECTRODES

In SOFC electrodes (anodes and cathodes), electrochemical reactions occur mainly in TPB. In these electrodes, the ions migrate through the ionic phase and the electrons flow occurs at the electronic conducting phase [47, 48].

TPB needs to be connected to the rest of the electrode structure, pores should be connected through the pore network (containing fuel or air), the electron conductive phase (mostly Ni in Ni-GDC cermets) must be connected, the current collector and the ion conductive ceramic phase (YSZ) or ions and electrons (GDC in reducing conditions of the anode) must be connected to the ceramic electrolyte (YSZ or GDC). TPB length depends on the particle diameter. While a reduction in particle diameter increases the TPB length, the gas permeability is reduced due to the consequent decrease in electrode porosity. When the particle diameter is about 1 μm, most of the electrochemical reaction runs at 10 μm from the electrolyte/anode interface and at 50 μm from the electrolyte/cathode interface [49].

In electrodic processes with current flow, the total current i_T is:

$$i_T = i_a + i_c \tag{1}$$

Where i_a and i_c are the anodic and cathodic current, respectively.

In equilibrium, $i_T = 0$ and $|i_a| = |i_c| = i_0$, where i_0 is the exchange current density. Electrode polarization is exhibited when electrode (E) and equilibrium ($E_{equilibrium}$) potential becomes different. Polarization measurement is called overpotential (η):

$$\eta = E - E_{equilibrium} \tag{2}$$

In electrochemical kinetics, it is fundamental to determining how the current density i varies with the overpotential η, then we have $i = f(\eta)$ or $i = f(E)$. The total overpotential (η_{total}) of a SOFC is the sum of several overpotentials contributions, each overpotential is related to a physical-chemical phenomenon caused by the current passage in the electrode/electrolyte interface.

$$\eta_{total} = \eta_A + \eta_C + \eta_R \tag{3}$$

Where:

η_A = Activation polarization, characterized by the charge transfer resistance at the electrode interface.

η_C = Diffusion polarization (mass transfer or concentration), characterized by the diffusion of the species resistance involved in the reaction to the electrode interface.

η_R = Ohmic polarization characterized by the ohmic resistance of the cell components.

The activation polarization is an energy barrier created for the transfer of electrons in the electrode interface. The electron transfer step may be determinant of reaction. η_A is given by the Buttler-Volmer, we can assume that there is only a single electron transfer at the interface.

$$i = i_0 \{ e^{-(F\beta \, \eta A/RT)} - e^{-(F(1-\beta) \, \eta A/RT)} \} \tag{4}$$

Where F is the Faraday constant and β is a dimensionless symmetry factor ranging between 0 and 1, with typically values close to 0.5. Physically we can consider that the activation energy barrier is the result of the rupture of bonds (chemistry, Van der Waals *etc.*) that occur when electroactive species begins its trajectory toward the electrode. In SOFC electrodes, the activation polarization is influenced by the number of reactive sites (TPB), which is mainly affected by the microstructure of the electrodes.

For mass transfer polarization, the reaction rate is determined by arrival of the reactant species to the active surface electrode. The concentration polarization is related to the mass transport resistance in the electrodes, usually neglected when the electrode thickness is less than 50 µm due to high gas diffusion rate at high temperatures. Nevertheless, studies of thick anodes, such as shown in Shi *et al.* [104], with a thickness greater than 600 µm, this type of polarization must be considered [50, 51].

Ohmic overpotential η_R is given by law of Ohm, considering R as electrolyte resistance. Ohmic overpotential occurs when the electrolyte conductivity is low, the concentration of the reagent is small or when very high currents exist.

$$\eta_R = RI \qquad \qquad (5)$$

Experimentally, the performance of a fuel cell is given by its polarization curve characteristics, as shown in Fig. (**2**). This curve presents single cell potential decrease in open circuit in current density (i) function. In low current densities η_R and η_C are negligible, and η_A mostly acts in the system behavior which means that the main overpotential comes from the resistance to charge transfer at the electrode surface, furthermore, this particular range is represented by the Buttler-Volmer equation (Eq. 4). For high current densities, electrolyte resistivity becomes predominant as well, and the system behaves as an ohmic resistance. When the system has very high current densities, the system is controlled by concentration potential [52].

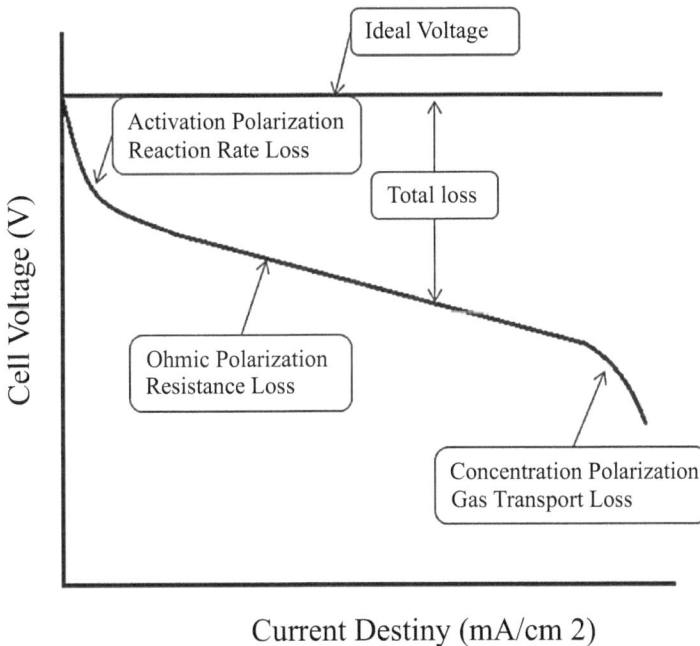

Fig. (2). Typical behavior of SOFC overpotential.

In SOFC anodes, the hydrogen oxidation reaction involves many chemical species and several steps, such as H_2 adsorption/desorption, surface diffusion and charge transfer, which is more complex than the oxygen reduction reaction at the cathode [53]. The first step of the reaction is the dissociative adsorption of H_2 on the Ni particles surface ($2H_{ad, Ni}$) (Eq. 6). The next reaction step is the charge transfer to produce $H^+_{ad,Ni}$ protons (Eq. 7) which diffuse into the reaction site in Ni/electrolyte interface (Eq. 8). The last step is characterized by the reaction of the protons with oxygen ions in the electrolyte, forming hydroxyl ions OH⁻ on the electrolyte surface (Eq. 9), and water desorption (Eq. 10). Reaction mechanisms are illustrated in Fig. (3).

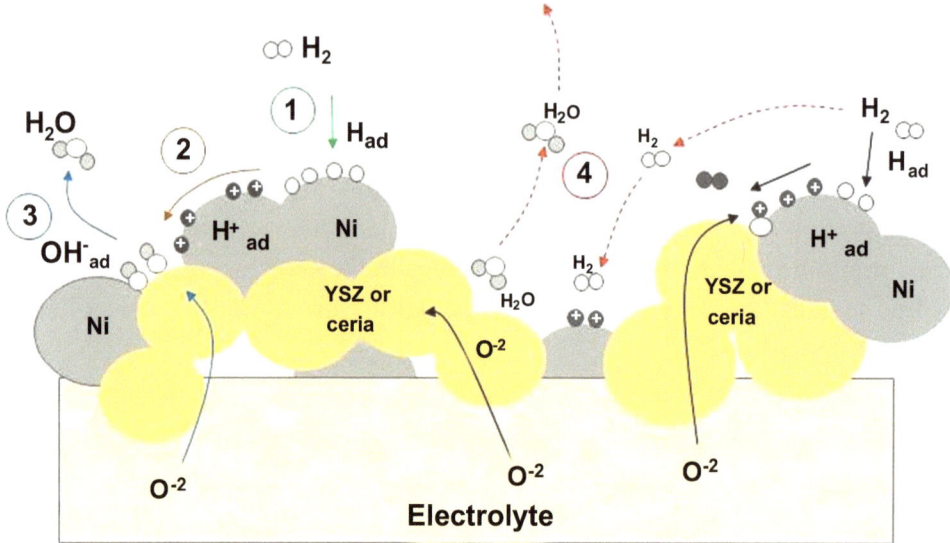

Fig. (3). Illustration of anodic reaction mechanisms taking place in Ni-(YSZ or Ceria) in H_2 atmosphere.

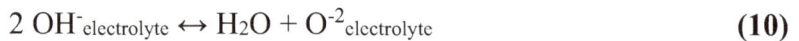

$$H_2 \leftrightarrow 2H_{ad, Ni} \tag{6}$$

$$2 \cdot (H_{ad, Ni} \leftrightarrow H^+_{ad,Ni} + e) \tag{7}$$

$$H^+_{ad,Ni} \rightarrow Ni/electrolyte\ interface \tag{8}$$

$$2 \cdot (H_{ad, Ni} + O^{-2}_{electrolyte} \leftrightarrow OH^-_{electrolyte}) \tag{9}$$

$$2\ OH^-_{electrolyte} \leftrightarrow H_2O + O^{-2}_{electrolyte} \tag{10}$$

1. Dissociative adsorption of hydrogen on the Ni surface and charge transfer to produce H^+.
2. Protons surface diffusion to the reaction sites in Ni/electrolyte interface.
3. Formation of OH⁻ and desorption of water.
4. Hydrogen diffusion into the pores and water out of the pores.

Although these mechanisms are very complex, they are not able to explain all experimental observations for the reaction in an anode of a SOFC, since additional reactions, such as oxidation of Ni to NiO, can play a key role in the overall reaction. For purposes of simplification, it is important to know that the complexity of the hydrogen oxidation reaction at the anode involves charge transfer steps and diffusion, which can be monitored by impedance spectroscopy measurements. The diffusion process may be limited by the difficulty of transporting H_2 and/or transporting reaction product (H_2O) and adsorbed species at the anode [53].

CONCLUDING REMARKS

Solid Oxide Fuel Cells (SOFC) offer a good alternative for the clean energy generation. Regarding the development of SOFC electrode materials, the most important properties are their catalytic activity for reduction and oxidation reactions and the compatibility with the electrolyte. The microstructure of the SOFC electrodes can be optimized *via* several processing routes and fabrication process.

Although lanthanum manganite-based materials are not the best ones for SOFC cathodes, mainly because of their low electrochemical activity, these are the most widely perovskites used with YSZ electrolyte because of their good chemical compatibility. Other materials like Sr and Co, Cu-doped lanthanum ferrites have been broadly studied for intermediate temperature SOFC.

Ni/YSZ is the most preferred anode materials for SOFC, however its inherent drawbacks such as low conductivity poses a great concern for its practical applications at low temperatures. Ni/Gd doped CeO_2 is a viable anode for low temperatures SOFC due to its remarkable high electrocatalytic activity at low temperatures, allowing the direct oxidation of hydrocarbon gases. Regardless of the materials used as SOFC electrodes, the high performance of electrodes depends on its microstructures.

CONFLICT OF INTEREST

The authors confirm that they have no conflict of interest to declare for this publication.

ACKNOWLEDGEMENTS

The authors acknowledgment CAPES (Coordenação de Aperfeiçoamento de Pessoal de Nível Superior), CNPq (Conselho Nacional de Desenvolvimento Científico e Tecnológico) and PPGCEM-UFRN (Programa de Pós-Graduação em Ciência e Engenharia de Materiais da Universidade Federal do Rio Grande do Norte) for the financial support.

REFERENCES

[1]   Atkinson, A.; Barnett, S.; Gorte, R.J.; Irvine, J.T.; McEvoy, A.J.; Mogensen, M.; Singhal, S.C.; Vohs, J. Advanced anodes for high-temperature fuel cells. *Nat. Mater.*, **2004**, *3*(1), 17-27.
[http://dx.doi.org/10.1038/nmat1040] [PMID: 14704781]

[2]   Liu, J.; Barnett, S.A. Operation of anode-supported solid oxide fuel cells on methane and natural gas. *Solid State Ion.*, **2003**, *158*, 11-16.
[http://dx.doi.org/10.1016/S0167-2738(02)00769-5]

[3]   Holzer, L.; Münch, B.; Iwanschitz, B.; Cantoni, M.; Hocker, T.; Graule, T. Quantitative relationships between composition, particle size, triple phase boundary length and surface area in nickel-cermet anodes for Solid Oxide Fuel Cells. *J. Power Sources*, **2011**, *196*, 7076-7089.
[http://dx.doi.org/10.1016/j.jpowsour.2010.08.006]

[4]   Mizusaki, J.; Tagawa, H.; Saito, T.; Yamamura, T.; Kamitani, K.; Hirano, K. Kinetic studies of the reaction at the nickel pattern electrode on YSZ in H_2-H_2O atmospheres. *Solid State Ion.*, **1994**, *70-71*, 52-58.
[http://dx.doi.org/10.1016/0167-2738(94)90286-0]

[5]   Zhu, W.Z.; Deevi, S.C. A review on the status of anode materials for solid oxide fuel Cells. *Mater. Sci. Eng.*, **2003**, *362*, 228-239.
[http://dx.doi.org/10.1016/S0921-5093(03)00620-8]

[6]   Haslam, JJ; Pham, AQ; Chung, B; Dicarlo, JF; Glass, RT Effects of the Use of Pore Formers on Performance of an Anode Supported Solid Oxide Fuel Cell. *J Am Ceram Soc.*, **2005**, *88*, 513-518.

[7]   Clemmer, R.M.; Corbin, S.F. Influence of porous composite microstructure on the processing and properties of solid oxide fuel cell anodes. *Solid State Ion.*, **2004**, *166*, 251-259.
[http://dx.doi.org/10.1016/j.ssi.2003.12.009]

[8]   Hu, J.; Lü, Z.; Chen, K.; Huang, X.; Ai, N.; Du, X.; Fu, C.; Wang, J.; Su, W. Effect of composite pore-former on the fabrication and performance of anode-supported membranes for SOFCs. *J. Membr. Sci.*, **2008**, *318*, 445-451.
[http://dx.doi.org/10.1016/j.memsci.2008.03.008]

[9]   Gil, V.; Moure, C.; Tartaj, J. Chemical and thermomechanical compatibility between Ni-GDC anode and electrolytes based on ceria. *Ceram. Int.*, **2009**, *35*, 839-846.
[http://dx.doi.org/10.1016/j.ceramint.2008.03.004]

[10]   Simner, S.P.; Anderson, M.D.; Pederson, L.R.; Stevenson, J.W. Performance Variability of LaSrFeO3 SOFC Cathode with Pt, Ag, and Au Current Collectors. *J. Electrochem. Soc.*, **2005**, *152*, A1851-A1859.
[http://dx.doi.org/10.1149/1.1995687]

[11]   Jiang, S.P. *Science and Technology of Zirconia V*; Badwal, S.P.; Bannister, M.J.; Hannink, R.H., Eds.; Technomic Publishing Company, **1993**, p. 819.

[12]   Mizusaki, J.; Tagawa, H.; Isobe, K.; Tajika, M.; Koshiro, I.; Maruyama, H. Hirano. K. Kinetics of the Electrode-Reaction at the H_2-H_2O Porous Pt Stabilized zirconia Interface. *J. Electrochem. Soc.*, **1994**, *141*, 1674.
[http://dx.doi.org/10.1149/1.2054982]

[13]   Suzuki, M.; Sasaki, H.; Otoshi, S.; Kajimura, A.; Ippommatsu, M. High power density solid oxide electrolyte fuel cells using Ru/Y_2O_3 stabilized zirconia cermet anodes. *Solid State Ion.*, **1993**, *62*, 125.
[http://dx.doi.org/10.1016/0167-2738(93)90260-A]

[14]   Spacil, H.S. Electrical device including nickel-containing stabilized zirconia electrode. US Patent 3,503,809, 1970.

[15]   Gorte, R.J.; Park, S.; Vohs, J.M.; Wang, C. Anodes for direct oxidation of dry hydrocarbons in a solid oxide fuel cell. *Adv. Mater.,* **2000**, *12*, 1465-1469.
[http://dx.doi.org/10.1002/1521-4095(200010)12:19<1465::AID-ADMA1465>3.0.CO;2-9]

[16]   Brown, M.; Primdahl, S.; Mogesen, M. Structure/performance relations for Ni/yttria-stabilized zirconia anodes for solid oxide fuel cells. *J. Electrochem. Soc.,* **2000**, *147*, 475-485.
[http://dx.doi.org/10.1149/1.1393220]

[17]   Tanner, C.W.; Fung, K.Z.; Virkar, A.V. The effect of porous composite electrode structure on solid oxide fuel cell performance,1. Theoretical analysis. *J. Electrochem. Soc.,* **1997**, *144*, 21-30.
[http://dx.doi.org/10.1149/1.1837360]

[18]   Muecke, U.P.; Graf, S.; Rhyner, U.; Gauckler, L.J. Microstructure and electrical conductivity of nanocrystalline nickel- and nickel oxide/gadolinia-doped ceria thin films. *Acta Mater.,* **2008**, *56*, 677-687.
[http://dx.doi.org/10.1016/j.actamat.2007.09.023]

[19]   Jiang, S.P.; Chan, S.H. A review of anode materials development in solid oxide fuel cells. *J. Mater. Sci.,* **2004**, *39*, 4405-4439.
[http://dx.doi.org/10.1023/B:JMSC.0000034135.52164.6b]

[20]   Steele, B.C.; Heinzel, A. Materials for fuel-cell technologies. *Nature,* **2001**, *414*(6861), 345-352.
[http://dx.doi.org/10.1038/35104620] [PMID: 11713541]

[21]   Belyaev, V.D.; Politova, T.I.; Marina, O.A.; Sobyanin, V.A. Internal steam reforming of methane over Ni-based electrode in solid oxide fuel cells. *Appl Catal A,* **1995**, *133*, 47-57.
[http://dx.doi.org/10.1016/0926-860X(95)00184-0]

[22]   Ormerod, R.M. Internal Reforming in Solid Oxide Fuel Cells. *Stud. Surf. Sci. Catal.,* **1999**, *122*, 35-46.
[http://dx.doi.org/10.1016/S0167-2991(99)80131-1]

[23]   Cheng, J.; Deng, L.; Zhang, B.; Shi, P.; Meng, G. Properties and microstructure of NiO/SDC materials for SOFC anode applications. *Rare Met.,* **2007**, *26*, 110-117.
[http://dx.doi.org/10.1016/S1001-0521(07)60169-7]

[24]   Janardhanan, V.M.; Heuveline, V.; Deutschmann, O. Three phase boundary length in solid-oxide fuel cells: A mathematical model. *J. Power Sources,* **2008**, *178*, 368-372.
[http://dx.doi.org/10.1016/j.jpowsour.2007.11.083]

[25]   Ding, C.; Lin, H.; Sato, K.; Hashida, T. Synthesis of NiO–$Ce_{0.9}Gd_{0.1}O_{1.95}$ nanocomposite powders for low-temperature solid oxide fuel cell anodes by co-precipitation. *Scr. Mater.,* **2009**, *60*, 254-256.
[http://dx.doi.org/10.1016/j.scriptamat.2008.10.020]

[26]   Deng, X.; Petric, A. Geometrical modeling of the triple-phase-boundary in solid oxide fuel cells. *J. Power Sources,* **2005**, *140*, 297-303.
[http://dx.doi.org/10.1016/j.jpowsour.2004.08.046]

[27]   Burmistrov, I.; Agarkov, D.; Tartakovskii, I.; Kharton, V.; Bredikhin, S. Performance Optimization of Cermet SOFC Anodes: An Evaluation of Nanostructured NiO. *ECS Trans.,* **2015**, *68*, 1265-1274.
[http://dx.doi.org/10.1149/06801.1265ecst]

[28]   Jiang, S.P.; Callus, P.J.; Badwal, S.P. Fabrication and performance of Ni/3 mol% Y_2O_3–ZrO_2 cermet anodes for solid oxide fuel cells. *Solid State Ion.,* **2000**, *132*, 1-14.
[http://dx.doi.org/10.1016/S0167-2738(00)00729-3]

[29]   Okawa, Y.; Hirata, Y. Sinterability, microstructures and electrical properties of Ni/Sm-doped ceria cermet processed with nanometer-sized particles. *J. Eur. Ceram. Soc.,* **2005**, *25*, 473-480.
[http://dx.doi.org/10.1016/j.jeurceramsoc.2004.01.011]

[30] Tamm, K.; Kivi, I.; Anderson, E.; Möller, P.; Nurk, G.; Lust, E. Optimization of Solid Oxide Fuel Cell Ni-CGO Anode Porosity. *ECS Trans.,* **2011**, *35*, 1771-1779.

[31] Chavan, A.U.; Jadhav, L.D.; Jamale, A.P.; Patil, S.P.; Bhosale, C.H.; Bharadwaj, S.R.; Patil, P.S. Effect of variation of NiO on properties of NiO/GDC (gadolinium doped ceria) nano-composites. *Ceram. Int.,* **2012**, *38*, 3191-3196.
[http://dx.doi.org/10.1016/j.ceramint.2011.12.023]

[32] Zha, S.; Rauch, W.; Liu, M. Ni-Ce$_{0.9}$Gd$_{0.1}$O$_{1.95}$ anode for GDC electrolyte-based low temperature SOFCs. *Solid State Ion.,* **2004**, *166*, 241-250.
[http://dx.doi.org/10.1016/j.ssi.2003.11.012]

[33] Misono, T.; Murata, K.; Fukui, T.; Chaichanawong, J.; Sato, K.; Abe, H.; Naito, M. Ni-SDC cermet anode fabricated from NiO–SDC composite powder for intermediate temperature SOFC. *J. Power Sources,* **2006**, *157*, 754-757.
[http://dx.doi.org/10.1016/j.jpowsour.2006.01.074]

[34] Babaei, A.; Jiang, S.P.; Li, J. Electrocatalytic promotion of palladium nanoparticles on hydrogen oxidation on Ni/GDC anodes of SOFCs *via* Spillover. *Electrochem Soc,* **2009**, *156*, B1022-B1029.
[http://dx.doi.org/10.1149/1.3156637]

[35] Lee, Y.; Joo, J.H.; Choi, G.M. Effect of electrolyte thickness on the performance of anode-supported ceria cells. *Solid State Ion.,* **2010**, *181*, 1702-1706.
[http://dx.doi.org/10.1016/j.ssi.2010.09.020]

[36] Kim, P.; Brett, D.J.; Brandon, N.P. The effect of water content on the electrochemical impedance response and microstructure of Ni-CGO anodes for solid oxide fuel cells. *J. Power Sources,* **2009**, *189*, 1060-1065.
[http://dx.doi.org/10.1016/j.jpowsour.2008.12.150]

[37] Ishihara, T.; Shibayama, T.; Nishiguchi, H.; Takita, Y. Nickel–Gd-doped CeO$_2$ cermet anode for intermediate temperature operating solid oxide fuel cells using LaGaO$_3$-based perovskite electrolyte. *Solid State Ion.,* **2000**, *132*, 209-216.
[http://dx.doi.org/10.1016/S0167-2738(00)00660-3]

[38] Rösch, B.; Tu, H.; Stormer, A.O.; Müller, A.C.; Stimming, U. Electrochemical characterization of Ni-Ce0.9Gd0.1O2−δ for SOFC anodes. *Solid State Ion.,* **2004**, *175*, 113-117.
[http://dx.doi.org/10.1016/j.ssi.2004.09.022]

[39] Changjing, F.; Siew, C.H.; Qinglin, L.; Xiaoming, G.; Pasciak, G. Fabrication and evaluation of Ni-GDC composite anode prepared by aqueous-based tape casting method for low-temperature solid oxide fuel cell. *Int. J. Hydrogen Energy,* **2010**, *35*, 301-307.
[http://dx.doi.org/10.1016/j.ijhydene.2009.09.101]

[40] Gil, V.; Moure, C.; Tartaj, J. Sinterability, microstructures and electrical properties of Ni/Gd-doped ceria cermets used as anode materials for SOFCs. *J. Eur. Ceram. Soc.,* **2007**, *27*, 4205-4209.
[http://dx.doi.org/10.1016/j.jeurceramsoc.2007.02.119]

[41] Tsipis, V.; Kharton, V.V.; Bashmakov, I.A.; Naumovich, E.N.; Frade, J.R. Cellulose-precursor synthesis of nanocrystalline Ce$_{0.8}$Gd$_{0.2}$O$_{2-\delta}$ for SOFC anodes. *J. Solid State Electrochem.,* **2004**, *8*, 674-680.
[http://dx.doi.org/10.1007/s10008-004-0507-z]

[42] Chen, J.C.; Hwang, B.H. Microstructure and properties of the Ni–CGO composite anodes prepared by the electrostatic-assisted ultrasonic spray pyrolysis method. *J. Am. Ceram. Soc.,* **2008**, *91*, 97-102.
[http://dx.doi.org/10.1111/j.1551-2916.2007.02109.x]

[43]   Ding, C.; Lin, H.; Sato, K.; Kawada, T.; Mizusaki, J.; Hashida, H. Improvement of electrochemical performance of anode-supported SOFCs by NiO- $Ce_{0.9}Gd_{0.1}O_{1.95}$ nanocomposite powders. *Solid State Ion.,* **2010**, *181*, 1238-1243.
[http://dx.doi.org/10.1016/j.ssi.2010.06.037]

[44]   Ding, C.; Sato, K.; Mizusaki, J.; Hashida, T. A comparative study of $NiO–Ce_{0.9}Gd_{0.1}O_{1.95}$ nanocomposite powders synthesized by hydroxide and oxalate co-precipitation methods. *Ceram. Int.,* **2012**, *38*, 85-92.
[http://dx.doi.org/10.1016/j.ceramint.2011.06.041]

[45]   Cela, B.; Macedo, D.A.; Souza, G.L.; Martinelli, A.E.; Nascimento, R.M.; Paskocimas, C.A. NiO–CGO *in situ* nanocomposite attainment: one step synthesis. *J. Power Sources,* **2011**, *196*, 2539-2544.
[http://dx.doi.org/10.1016/j.jpowsour.2010.11.026]

[46]   Macedo, D.A.; Figueiredo, F.M.; Paskocimas, C.A.; Martinelli, A.E.; Nascimento, R.M.; Marques, F.M. Ni–CGO cermet anodes from nanocomposite powders: Microstructural and electrochemical assessment. *Ceram. Int.,* **2014**, *40*, 13105-13113.
[http://dx.doi.org/10.1016/j.ceramint.2014.05.010]

[47]   Lai, W.; Haile, S.M. Impedance spectroscopy as a tool for chemical and electrochemical analysis of mixed conductors: a case study of ceria. *J. Am. Ceram. Soc.,* **2005**, *88*, 2979-2997.
[http://dx.doi.org/10.1111/j.1551-2916.2005.00740.x]

[48]   Chueh, W.C.; Hao, Y.; Jung, W.; Haile, S.M. High electrochemical activity of the oxide phase in model ceria-Pt and ceria-Ni composite anodes. *Nat. Mater.,* **2011**, *11*(2), 155-161.
[http://dx.doi.org/10.1038/nmat3184] [PMID: 22138788]

[49]   Andersson, M.; Yuan, J.; Sundén, B. Review on modeling development for multiscale chemical reactions coupled transport phenomena in solid oxide fuel cells. *Appl. Energy,* **2010**, *87*, 1461-1476.
[http://dx.doi.org/10.1016/j.apenergy.2009.11.013]

[50]   Shi, Y.; Cai, N.; Li, C.; Bao, C.; Croiset, E.; Qian, J.; Hu, Q.; Wang, S. Modeling of an anode-supported Ni–YSZ|Ni–ScSZ|ScSZ|LSM–ScSZ multiple layers SOFC cell Part I.Experiments, model development and validation. *J. Power Sources,* **2007**, *172*, 235 245.
[http://dx.doi.org/10.1016/j.jpowsour.2007.04.037]

[51]   Yang, S.; Chen, T.; Wang, Y.; Peng, Z.; Wang, W.G. Electrochemical analysis of an anode-supported SOFC. *Int. J. Electrochem. Sci.,* **2013**, *8*, 2330-2344.

[52]   Linardi, M. *Introdução à ciência e tecnologia de células a combustível*; Artliber Editora Ltda, **2010**.

[53]   Gödickemeier, M. Mixed ionic electronic conductors for solid oxide fuel cells, 287f. Dissertation ETH Swiss Federal Institute of Technology, Zurich, **1996**, 11348.

Recent Advances in Synthesis of Lanthanum Silicate Apatite Powders as New Oxygen-Ion Conductor for IT-SOFCs: A Review

Chieko Yamagata[*], **Daniel R. Elias**, **Agatha M. Misso** and **Fernando S. Santos**

Nuclear and Energy Research Institute, Department of Materials Science and Technology, Sao Paulo, Brazil

Abstract: YSZ (yttrium stabilized zirconia) with fluorite structure is the traditional electrolyte used in SOFC (solid oxide fuel cell), where the operating temperatures are above 900 °C. In those high temperatures, reactions between the components may occur, in addition to the thermal expansion and contraction, causing the diminution of the cell life. Therefore, the reducing of operating temperature to development of the intermediate temperature SOFs (IT-SOFCs) is one of important task to the power production area. Find an alternative electrolyte for IT-SOFCs became the interest of researchers. Recently, rare earth silicate-based compositions materials with apatite-type structure, with general formula $Ln_{10-a}Si_6O_{26+b}$, (where Ln is La, Sm, Nd, Dy or Gd, and a = 8 to 11), have attracted significant attention as electrolyte. This is because of the structure allows high ionic conductivity with low activation energy at intermediate temperatures. For example, the lanthanum silicate apatite (LSA) solid electrolyte, with the composition $La_{10}Si_6O_{27}$ has exhibited high oxygen ionic conductivity at 500 °C. However, several problems in obtaining the pure apatite single phase and the low sinterability of LSA are disadvantageous for its application as electrolyte. Therefore, the development of a viable synthesis process to attain LSA apatite crystalline powder, with high sinterability for applying in the production of IT-SOFCs electrolytes, turned out to be a challenge for SOFC researchers. Efforts have focused to reach the pure single phase of apatite with the reduction of temperature and time of the sintering process. In this review, different methods, of LSA synthesis, are summarized and discussed.

Keywords: Apatite, Characterization, Electrolyte, Ionic conductivity, $La_{10}Si_6O_{27}$, Lanthanum silicate, Methods comparison, SOFC, Synthesis, X-ray Diffraction.

---

[*] **Corresponding author Chieko Yamagata:** Nuclear and Energy Research Institute, Department of Materials Science and Technology, Sao Paulo, Brazil; Tel: +55-11-31339389; Fax: +55-11-31339276; E-mail: yamagata@ipen.br

Moisés R. Cesário & Daniel A. de Macedo (Eds.)

INTRODUCTION

The main challenge in the development of fuel cell technology involves the production of the component layers of SOFC (solid oxide fuel cell) cell stack. High ionic conductivity in addition to low electronic conductivity at operation temperature [1] and be secure in equally oxidizing and in reducing environments is essential for electrolyte of SOFC. The mechanism of the flow current in the electrolyte consists on the oxygen ions in movement through the crystal lattice. The moving of thermally activated ions, from one crystal lattice to its neighbor site [2], results the current. This mechanism occurs in stabilized zirconia in SOFC electrolyte, where the oxygen vacancies created by reducing of Zr^{4+} ion in the lattice site, establishes the ion oxygen conductivity. The operating temperature of SOFC must be satisfactory to permit achieving the ionic conductivity into the electrolyte. Typical SOFC electrolyte is yttria doped zirconia with 8 mol% (8YSZ), in which the electronic conductivity is negligible even under highly reducing environments [1]. Other electrolytes under considering are Ga-doped ceria (GDC), Sm-doped ceria (SDC), Sc-stabilized zirconia (SDZ) and La-Sr--a-Mg oxide (LSGM), because of their high ionic conductivities at reduced temperatures [3].

YSZ, the most important electrolyte used as SOFCs electrolyte, exhibits the fluorite structures and has high mechanical strength. However, it requires high temperatures (800 – 1000 °C) in order to obtain its efficient performance. Working at this range of temperatures reduces the useful life of SOFC and limits the selection of materials. Therefore, the main feature of the SOFC researches focuses in the development of high ionic conductivity solid electrolytes at reduced temperatures. It is one of the important tasks to overcome for the development of named intermediate temperature solid oxide fuel cells (IT-SOFC), to operate at temperatures between 500 and 750 °C [2].

As CeO_2 and $La_{0.9}Sr_{0.1}Ga_{0.8}Mg_{0.2}O_{3-x}$ (LSGM) exhibit higher ionic conductivity than YSZ [4] at reduced temperatures, they may be considered as alternative electrolyte for IT-SOFC. However, they develop electronic conduction under reducing environments.

The high electric conductivity of apatite-type lanthanum silicate has recently attracted significant interest of the researchers. Nakayama *et al.* [5, 6] first reported that lanthanum silicates ($Ln_{10-a}Si_6O_{26+b}$ (Ln = La, Sm, Nd, Dy, Gd, a = 8 to 11) with hexagonal crystal system similar to calcium phosphate $Ca_{10}(PO_4)_6O_2$ called apatite, exhibited ionic conductivity. In their studies, one with the composition $La_{10}Si_6O_{27}$ has exhibited oxygen ionic conductivity of 10^{-3} Scm^{-1} at

500 °C, which is higher than that of YSZ electrolytes [5, 7] at that temperature. Lanthanum silicate has hexagonal crystal system similar to calcium phosphate $Ca_{10}(PO_4)_6O_2$. Later, detailed investigation [8 - 10] of crystal structure and conduction mechanism of this oxide electrolyte revealed that the conductivity in apatite systems involves interstitial oxide ions in addition to oxygen vacancies mechanism. Various studies performed by theoretical atomistic simulations [4, 10 - 13], neutron powder diffraction [14 - 18] and Mössbauer [16] spectroscopies, have recognized the conductivity within the apatite structure occurs by interstitial migration of oxide anions. Interstitial oxide dominates ionic conductivity of lanthanum silicate and pass in a *c*-axis direction [2, 3]. By doping other elements into sites in the crystals, the ionic conductivity may be improved. The conductivity of lanthanum silicate apatite (LSA) increases with La content in $La_{9.33+x}Si_6O_{26+1.5x}$ because of excess oxygen ions introduced into the interstitial position of lattice [19]. $La_{9.92}Si_6O_{26.88}$ shows higher conductivity than YSZ below 650 °C [19]. Through lattice doping, the conductivity of LSA can further improved. The conductivity of oxygen ions can be enhanced to a value of 88 mS.cm^{-1}, with activation energy as low as 0.42 eV at 800 °C, by a small Mg substitution amount, in the Si site [20]. This conductivity is higher than YSZ below 900 °C and similar to doped $LaGaO_3$ below 550 °C. Doping Al and Ba into Si site and La site respectively is reported to enhance the conductivity [21, 22], as well as excess oxide ion and vacancy in La site contribute to the improvement of ionic conductivity in the LSA [23, 24]. Therefore, LSA is a potential candidate as a new electrolyte. Materials for use as electrolyte are required to be dense [22, 25, 26]. The difficulty in obtaining dense ceramic of LSA is the main trouble for its application in SOFC. For example, to achieve ceramic bodies with relative densities higher than 90%, temperatures as higher as 1650 °C is need for sintering process using LSA powders synthesized by solid-state reaction. Densities higher than 93% require grinding of the powders, hot pressing and sintering at high temperatures [23, 27]. Additional trouble of LSA consists on the difficulty to obtain the pure apatite single phase. $La_{9.33}Si_6O_{26}$ and La_2SiO_5 are the well-known crystalline phases found in the samples of $La_{10}Si_6O_{27}$ sintered below 1775 °C. Therefore, the major research efforts have been focused to reach the pure single phase of apatite and reduction of temperature and time of the sintering process. To obtain a homogeneous mixture of the precursor oxides, La_2O_3 and SiO_2, by conventional solid-state reaction, is difficult. Secondary crystalline phases, such as La_2SiO_5 and $La_2Si_2O_7$ are often associated to LSA. Using sol–gel technique, the formation of those secondary phases may avoid, because the process allows distribution of the components more homogeneously. Nevertheless, in this method, parameters of the processing, such as, the concentration and molar ratio

of Si and La precursors, the pH for the gel formation, time of gelling, *etc.* they must be well controlled. Recently, new knowledge are reported [28, 29] into obtaining oxy-apatite dense materials. Freeze-dried precursor method allowed obtaining of homogeneous nanopowders. Spark plasma sintering has used for sintering processing. High dense ceramics were obtained at relatively low temperatures, from this technique. Practically, all of the process parameters have a great influence on the characteristics and size of synthesized powders and consequently on the relative density of the ceramics obtained from those powders. It is recognized weak homogeneous agglomerated of nano powders are important factors to attain high density of ceramics. Usually, alkoxide process in organic solvents permits the obtaining of high reactive and weak agglomerated nano powders. Furthermore, co-precipitation and hydrothermal methods are used to synthesize LSA powders with good characteristics to achieve dense ceramic. Among several methods for LSA preparing, the process based on melting of precursor reagent salts is a suggested method. Therefore, the development of a viable synthesis procedure to attain single-phase crystalline lanthanum silicate apatite (LSA) nano powders with high sinterability for applying in the production of IT-SOFCs solid electrolyte turned out be a challenge for SOFC researchers. In this review, the recent advances in synthesis of LSA powders, as new oxygen-ion conductor, are summarized and discussed.

LANTHANUM SILICATE APATITE (LSA)

Recently, a rare-earth silicates material with apatite structure is a proposed alternative solid electrolyte for IT-SOFC. The interest in the rare earth silicates materials for SOFC electrolyte application has grown [30] after Nakayama *et al.* [5] studies that reported those silicates have presented high oxide ion conductivities. Around temperatures of 700^0C, 1.08×10^{-2} S.cm^{-1} ionic conductivity of pure lanthanum silicate is not high, but it may be improved at lower temperatures, due to a low activation energy that is of 0.64 eV [6]. The promising application of LSA as IT-SOFC electrolyte is justified by the pure ionic conduction with a transport number near to 1 over an extensive range of partial oxygen pressure [22]. The mechanism of LSA conductivity at intermediate temperatures [5] is exclusively ionic interstitial conduction [24, 23, 31], different to the vacancy mechanism that is common to the other oxide ion conductors, such as stabilized zirconia and doped lanthanum gallates. The crystal structure of LSA [32] with general formula $Ln_{10}(SiO_4)_6O_{2\pm y}$ (Ln is a rare earth) is alike apatite crystalline phase found in bones and teeth. Fig. (**1**) shows apatite system structure. It can be seen isolated SiO_4 tetrahedral forming two distinct oxide ion and La channels laying parallel to the *c*-axis. La cations occupy the smaller of these channels and the larger one is occupied by oxide ions.

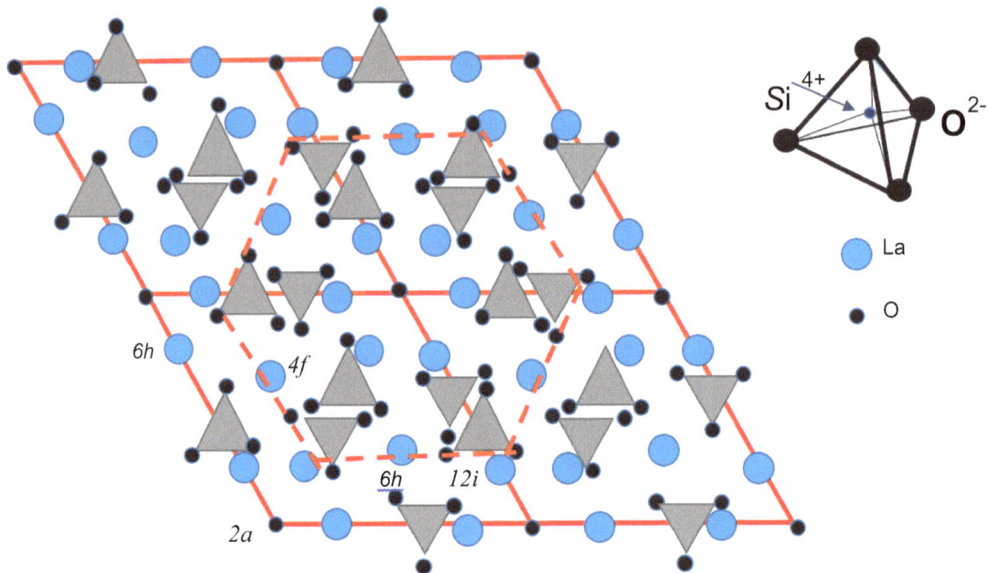

Fig. (1). The structure of La $_{9.33}$ Si $_6$ O $_{26}$ showing tetrahedral structure represented by SiO$_4$, large spheres by La and small spheres by O.

Studies of the conduction mechanism [33] are important for the optimization of these systems conductivity. Investigations have proved that apatite systems conductivity is not trivial due to the complex nature of its structure. It has been shown that an oxide-ion interstitial mechanism is involved [25] added to the initial suppositions that an oxide-ion vacancy mechanism. For these reason the oxide-ion conductivity of LSA is high. Results of doping studies have shown the importance of interstitial oxide-ions. The investigation revealed that high conductivity is favored by the presence of oxygen hyper stoichiometry (x>0 in the formula Ln$_{9.33+x}$(SiO$_4$)$_6$O$_{2+3x/2}$), or cation vacancies; the latter allowing the creation of interstitial oxide ions through promoting Frenkel-type disorder [24].

A detailed modelling investigation performed on the La$_{9.33}$(SiO$_4$)$_6$O$_2$ system by a computer modeling, it identified a interstitial position of oxide, near to the SiO$_4$ groups [12]. According to that study, the position is energetically favorable. This modelling work was later reinforced by structural studies of La $_{9.33+x}$ Si$_6$O $_{26+3x/2}$ systems, with excess of oxygen [34, 35]. The simulation of modelling also suggested that, in La$_{9.33}$(SiO$_4$)$_6$O$_2$ the conductivity occurs through interstitial approach, whereas in alkaline earth doped, La$_8$Sr$_2$(SiO$_4$)$_6$O$_2$, includes the vacancy way. The study has correlated this remark from the calculated and observed energies of activation [12]. The conduction by interstitial way consists on a complex sinusoidal procedure [36]. The high conductivities of the oxide ion [12, 34], as outlined is showed in Fig. (**2**). In order to observe high conductivity [12, 34], it is fundamental to have dislocations of the silicate structure, allowed by its flexibility.

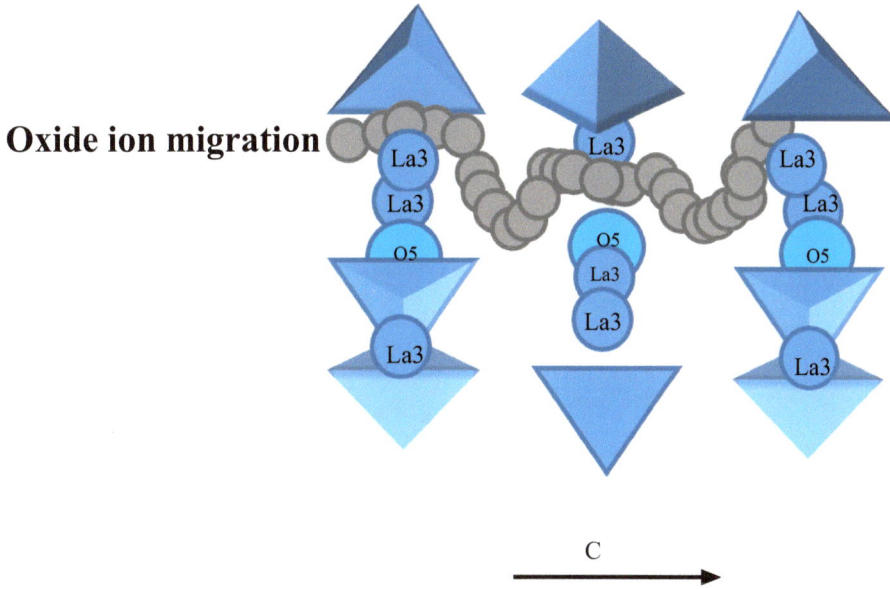

Fig. (2). Delineated interstitial mechanism of conductivity in $La_{9.33}Si_6O_{26,}$ observed, perpendicular to the oxide ion position.

Table **1** shows the bulk conductivities of rare earth silicates ($Ln_{9.33}Si_6O_{26}$, Ln = rare earth) without doping [36].

Table 1. [36]: Conductivity data of rare earth silicates with general formula $Ln_{9.33}Si_6O_{26}$.

Rare Earth	σ (Scm−1) at 500 °C	Ea (eV)
La	1.1×10^{-4}	0.74
Pr	8.1×10^{-5}	0.75
Nd	1.0×10^{-4}	0.72
Sm	2.2×10^{-5}	0.83
Eu	1.3×10^{-5}	0.85
Gd	1.5×10^{-6}	0.95

From Table **1**, it is observed that the conductivities of the rare earth La, Pr and Nd systems are around 1×10^{-4} eV. For the systems of Eu, Sm and Gd the conductivity decrease and the activation energy increase. It also observes that the conductivity depends on the element of rare earth associated in the silicate system. From Table **1**, the highest conductivity is observed for $La_{9.33}Si_6O_{26}$. Therefore, it has attracted research attention to its application as electrolyte in SOFC technology. One of the main features of LSA is the possibilities of enhancement of its conductivity by doping within the structure. For example, studies, of doping Al and Ba into Si site and La site respectively, has reported that the doping enhances the LSA conductivity [22]. In addition, excess oxide ion and vacancy in La site contribute to improvement of ionic conductivity in the lanthanum silicate [22, 23, 37].

LANTHANUM SILICATE APATITE (LSA) POWDER SYNTHESIS

Many techniques have been applied to prepare apatite-type materials. The traditional high temperature (> 1350 °C) solid-state reaction is the most usual route. Nevertheless, preparing LSA at a lower temperature, in a shorter time, and with high reproducibility would be strategic.

Among all apatites, lanthanum silicate (LSA) is an important promising candidate to an actual technical application. Many efforts have been made to synthesize LSA powders with higher sinterability, since it is difficult to prepare dense ceramic with pure apatite crystalline phase. With the purpose to overcome LSA inherent difficulties, alternative routes of synthesis are investigated, such as, coprecipitation, Pechini method, hydrothermal treatment, freeze drying, high energy ball milling, and plasma spraying. Recently, a novel modified route to synthesize LSA was proposed. The method joints two techniques: the sol-gel and precipitation. In this review, the recent advances in synthesis of LSA powders as new oxygen-ion conductor are summarized and discussed.

Conventional Solid-State Reaction Synthesis

Conventional solid-state reaction method has been the most widely used technique to synthesize oxide materials. It consists on the high temperature reaction in the mixture of precursor powders. The reaction is controlled by the diffusion of atomic or ionic species through the mixture. The high temperature and small size of the reagents are essential to solid-state method. The temperature required must be sufficient to provide high mobility of the reagent species. Small size of the particles allows them close together for the reaction occurring.

LSA is usually prepared by conventional solid-state synthesis route [22, 25, 29, 37 - 41]. There are many inconveniences in this method. Adding to requirement of a high temperature (1600–1800 °C) [22], an intermediate wet milling and long-time of sintering are need to achieve dense and single-phase ceramic [22, 39]. Other disadvantage of the method is the difficulty to control the composition, homogeneity, crystallinity, and grain size growth. Although same high quality samples should prepare successfully by this method, secondary crystalline phases, such as $LaSi_2O_5$ and $La_2Si_2O_7$ [37] are frequently formed. Dependent of those phase intensities the conductivity may decline.

Kendrick *et al.* [38] have prepared LSA single phase $La_{9.33}Si_6O_{26}$, La_2O_3 and SiO_2 source materials were dried and mixed at stoichiometry quantities, ground and twice thermal treated for 12h at 1350 °C. X-ray diffraction (XRD) analysis confirmed the phase purity of the powder. By submitting the mixed powder in NH_3 environment and temperature of 950 °C for 20 h, resulted the powder to study its reaction in these settings. The work suggest at 600 °C, there is very little nitridation, and so the use of LSA electrolytes in SOFCs in which NH_3 is the fuel, should operates at such lower temperatures. However, highlight the need for more fundamental studies about this question [38].

Solid state technique was used to obtain $La_{9.33}(Si_5V_1)O_{26}$ ceramics [42].Vanadium element was added because it melts at low temperature and has high electrical conductivity in the region between middle and low temperature. Those ceramics structural and electrical properties were investigated for the use in SOFC. According to the composition formula $La_{9.33}(Si_5V_1)O_{26}$, stoichiometric amounts of the starting materials La_2O_3, SiO_2 and V_2O_5 were mixed and ground in ethyl alcohol for 24h, using zirconia ball. Then, the mixture was thermal treated at 1200 °C in air, during 2h. Later, 3 wt% PVA was added to the blend and dried. Subsequently, pellets attained from those powders by 150 MPa hydrostatic pressing, were sintering at 1300~1450 °C for 2 h in air. The sample sintered at 1400°C is shown homogeneous, presented a maximum density of 4.93g.cm^{-3} and single apatite phase, verified by the XRD pattern. The density, of the samples did not increase even sintered at temperatures above 1450°C. It was explained by the effect of compensating, *i.e.*, the grain size increases and occurs the pores formation because V ion volatilizes.

Solid-state reaction process [43] was used to synthesize co-doped LSAs, $La_9.5X_{0.5}Si_{5.5}Fe_{0.5}O_{26.5}$ (X = Ca, Sr and Ba). Stoichiometric mass of raw materials (La_2O_3, SiO_2, $BaCO_3$, $SrCO_3$, $CaCO_3$ and $Fe(NO_3)_3.9H_2O$) were mixed in plastic recipient with isopropanol medium and zirconia balls, for 24 h to obtain compositions of $La_{10}Si_{6.0}O_{26.5}$, $La_{10}Si_5FeO_{26.5}$, $La_9.5Ba_{0.5}Si_{5.5}Fe_{0.5}O_{26.5}$, $La_9.5Sr_{0.5}Si_{5.5}Fe_{0.5}O_{26.5}$ and $La_9.5Ca_{0.5}Si_{5.5}Fe_{0.5}O_{26.5}$. In all samples, which were dried and after 10h calcined at 1300 °C, La_2SiO_5 secondary phase was observed. Ceramic bodies of

$La_{10}Si_{6.0}O_{26.5}$, $La_{10}Si_5FeO_{26.5}$, $La_{9.5}Sr_{0.5}Si_{5.5}Fe_{0.5}O_{26.5}$ and $La_{9.5}Ca_{0.5}Si_{5.5}Fe_{0.5}O_{26.5}$ were obtained from the calcined powders by sintering for 4h at 1550 °C. The SEM surface micrographs of sintered $La_{10}Si_{6.0}O_{26.5}$ sample showed very porous structure; however, $La_{10}Si_5FeO_{26.5}$ sample showed much denser structures with less porosity. It has observed that the grain size increases with sintering temperature increasing. At the sintering temperature of 1600 °C, the microstructure of $La_{9.5}Ba_{0.5}Si_{5.5}Fe_{0.5}O_{26.5}$ showed a denser structure with no porosity and large grains. The study concluded the densification process of $La_{9.5}Ba_{0.5}Si_{5.5}Fe_{0.5}O_{26.5}$ strongly depends on the sintering temperature.

To investigate the influence of Mg adding on the crystalline structure of silicate system with general formula $La_{10}Si_{6-x}Mg_xO_{27-x}$ (x = 0 to 0.4), and its conductivity, the solid-state process was used to prepare the ceramic precursor powder [44]. Lanthanum silicates $La_{10}Si_{6-x}Mg_xO_{27-x}$ (x = 0 to 0.4) was synthesized from La_2O_3, SiO_2, and MgO staring materials. After drying the raw materials for 2h at 1000 °C, they were ground for 8h at of 400 rpm in ethanol, followed by drying again and sifting. Afterwards, they were pressed to pellets and sintered at 1350 °C in the period of 10h. Finally, these pellets were reground, pressed and thermal treated at 1750 °C for 6 h. Well-crystallized apatite without impurities phases has verified for all samples, by comparison their XRD patterns to the JCPDS 53-0291.

A high pure crystalline apatite of $La_{10}Si_6O_{27}$ ceramic was obtained from powder synthesized *via* the high temperature solid-state reaction, sintered at 1923 K for 10h [45].

By heating at 1673 K, Kobayashi *et al.* [46] used two pairs of raw materials; $La(OH)_3$ and amorphous-SiO_2, and $La(OH)_3$ and quartz-SiO_2, to synthesize the silicate oxyapatite (LSO) ceramics single-phase powder. The use of lanthanum hydroxide was justified to simplify the process of synthesis and allowing the use of water as mixing environment. By this method, dense (>94%) LSO ceramics were obtained, at 1773 K. Nevertheless, in some cases, were verified impurity phases, resulted from the ball and pod contamination at milling process.

Pons *et al.* [47] used a new reagent $La_2O_2CO_3$, obtained by calcining La_2O_3 at 900 °C for 2h under CO_2 atmosphere. Stoichiometric amounts of the precursors, the new reagent and amorphous SiO_2 were used to synthesize LSA $La_{9.33+x}(SiO_4)_6O_{2+3x/2}$ $_{(-0.2,x,0.27)}$ by solid-state method, according to the reaction of the process (1):

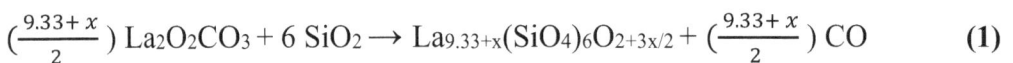

$$(\frac{9.33+x}{2}) \, La_2O_2CO_3 + 6 \, SiO_2 \rightarrow La_{9.33+x}(SiO_4)_6O_{2+3x/2} + (\frac{9.33+x}{2}) \, CO \qquad (1)$$

Compositions of $La_{9.13}(SiO_4)_6O_{1.7}$; $La_{9.33}(SiO_4)_6O_2$; $La_{9.53}(SiO_4)_6O_{2.3}$ and $La_{9.60}(SiO_4)_6O_{2.405}$ were synthetized by calcining the mixture of precursors for 4h at temperatures between 1000 and 1500 °C. The formation of $La_2Si_2O_7$ secondary phase was avoided using $La_2O_2CO_3$ instead La_2O_3. From powders pressed isostacally thermal treated at 1200 °C flowed by sintering at 1550 °C, dense ceramic body were attained.

Sol-gel Synthesis

The sol-gel [48, 49] method allows achieving crystalline or amorphous bulk materials, as well as films and fiber structured particles. It is useful to produce hybrid organic/inorganic nanoparticles, when the good dispersion of the nanophase is important to the final material. In this method, parameters of the reaction can be controlled in the homogeneous molecular media. Therefore, synthesis of uniform well-defined morphology nanoparticles with high purity and homogeneity [50] can be reach. However, in despite of this, the sol-gel method is still little used compared to solid-state method, because of the complexity of the sol-gel chemistry [51]. There is a variety of parameters to control rigorously, such as pH, temperature, the nature and concentration of anions, method of mixing, rate of the metal oxide precursors *etc.*, making the controlling not relatively easy. Other difficulty of this method is that as-synthesized product is amorphous, requiring thermal treatment to its crystallization. Sometimes, this treatment may cause damages in the morphology, shape and size of particles obtained in the synthesis process. The sol-gel process consists on the preparation of a suitable media to the sol formation and successive gelation and solvent elimination. Tetraethyl silicate (or tetraethoxy-silane) $Si(OC_2H_5)_4$, TEOS, is the usual precursor compound in the synthesis of silicates. The sol-gel synthesis route was used successfully to prepare $La_{9.33}Si_6O_{26}$ ceramic powders [52, 53] with better homogeneity and purity than the solid-state powder synthesized. In addition, the optimization of the synthesis parameters and reactants, the temperature of apatite crystallization was reduced substantially. Consequently, the sintering temperature to obtain the ceramic can also diminished below that used by conventional solid-state synthesis.

Shi Qingle *et al.* [54] investigated the effect of pH value (from 1 to 5) of the solution on the homogeneity, characteristics and formation of the LSA sol. Lanthanum silicate with formula $La_{9.33}Si_6O_{26}$ was prepared by sol-gel process using TEOS and lanthanum nitrate as starting materials. They observed the viscosity of the sols slightly increases first and then increases abruptly because the predominant reaction mechanism changes, from hydrolysis reaction to condensation reaction. The onset time of the increase for the viscosity shortens from pH 1 to pH 4. The gelation time decreases with increasing pH of the solution. They suggest the pH less than 4 to the gel formation. The maximum value for the zeta potential and stability were observed at pH starting at 2. Lanthanum silicate obtained from this condition of pH presented the pure apatite crystalline phase. Others lanthanum silicates prepared from different initial pH showed impurity phases.

Using lanthanum nitrate or acetate and TEOS ($(OC_2H_5)_3Si\text{-}OC_2H_5$) as starting materials in ethanol solvent, Vojisavljevic´ *et al.* [55] have synthesized $La_{9.33}Si_6O_{26}$ by acid-catalyzed sol-gel process. TEOS was added to a solution of La (III) salts, where the reaction of hydrolysis (**2**), alcohol condensation (**3**) and water condensation (**4**) occur as flowing showed:

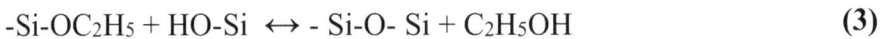

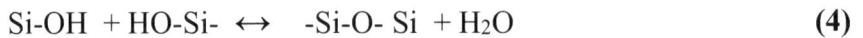

$$(OC_2H_5)_3Si\text{-}OC_2H_5 + H_2O \rightarrow (OC_2H_5)_3Si\text{-}OH + C_2H_5\text{-}OH \qquad \textbf{(2)}$$

$$-Si\text{-}OC_2H_5 + HO\text{-}Si \leftrightarrow -Si\text{-}O\text{-}Si + C_2H_5OH \qquad \textbf{(3)}$$

$$Si\text{-}OH + HO\text{-}Si\text{-} \leftrightarrow -Si\text{-}O\text{-}Si + H_2O \qquad \textbf{(4)}$$

Ethanol formed in the hydrolysis reaction in the (**3**) is important to the formation of a homogeneous gel. It allows the uniform distribution of TEOS in the liquid medium of reaction, since the TEOS is not miscible in water. The molar ratio of H_2O:Si as well as TEOS concentration effect on the reactions of hydrolysis and condensation and consequently the characteristics of the gel [52]. To avoid the crystalline phases that are considered impurities, such as, La_2O_3, La_2SiO_5 and $La_2Si_2O_7$, the molar ratio H_2O:Si must be low. The concentration of water in the medium of reaction depends on the lanthanum salt solubility. If the solubility of salt is high, the concentration of water is low. A high concentration of TEOS is suggested to achieve final powder with pure crystalline phase. The volumetric ratio value of the catalyzer:TEOS below 8 [52], not affect the obtained crystalline phases. Then, the optimized conditions are: ratio of hydrolysis and TEOS concentration, minimum and high, respectively.

Vojisavljevic´ and co-workers [55] reported the preparation of the $La_{9.33}Si_6O_{26}$ powder as showed in Fig. (**3**).

They obtained pure the $La_{9.33}Si_6O_{26}$ powder by the route of La-nitrate with the acetic acid as catalyzer, after calcining the precursor at 900 °C. Ceramic with pure crystalline phase and 94% relative density was resulted by sintering at 1400 °C. It is approximately 200 °C lower than the usual temperature reported to reach apatite crystalline phase.

```
┌────────────────────────────────────────────────────────────┐
│ Lanthanum nitrate (or lanthanum acetate) +Ethanol+acetic acid│
└────────────────────────────────────────────────────────────┘
                            │
                            ▼
        ┌──────────────────────────────────────┐
        │   Clear lanthanum acid solution        │
        └──────────────────────────────────────┘
                            │
                            ▼
        ┌──────────────────────────────────────┐
        │  TEOS (Si (OCH₂NO₃).6( H₂O)           │
        └──────────────────────────────────────┘
                            │  Stirring 1h at room temperature
                            ▼
            ┌──────────────────────────────┐
            │      Clear solution           │
            └──────────────────────────────┘
                            │  Heating at 60-70 °C in oil bath
                            ▼
            ┌──────────────────────────────┐
            │           Gel                 │
            └──────────────────────────────┘
                            │  Dry overnight at 80 °C
                            ▼
            ┌──────────────────────────────┐
            │         Xerogel               │
            └──────────────────────────────┘
                            │  Decomposition at 600-700 °C in air
                            ▼
        ┌──────────────────────────────────────┐
        │        Amorphous powder               │
        └──────────────────────────────────────┘
                            │  Calcination at 900-1300°C in air
                            ▼
        ┌──────────────────────────────────────┐
        │         Apatite powder                │
        └──────────────────────────────────────┘
```

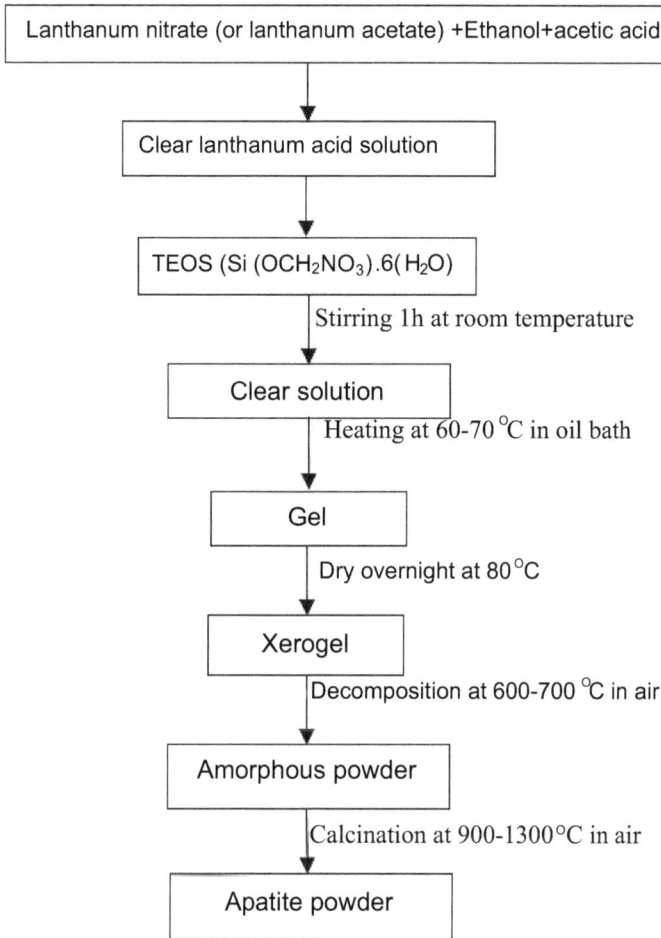

Fig. (3). The sol gel process flowchart, using the route of La-nitrate with the acetic-acid catalyst for synthesis of $La_{9.33}Si_6O_{26}$ lanthanum silicate powder.

On the other hand, from La-acetate precursor synthesis, poor homogeneity gel was formed. Impurities phases were detected in powders calcined at temperature above 900 °C.

Qin Li and Limin Dong [56] proposed the sol-gel aqueous route to prepare LSA, with formula $La_{9.33}Si_6O_{26}$. TEOS was added, with stirring, to a clear mixture solution of La_2O_3 in nitric acid, ethanol and acetic acid. After several hours, at 80 °C, the white gel formed. Apatite powder was obtained from this gel, after drying at 100 °C and calcinig at 1000 °C. By pressing the powder at high-pressure (4.5 GPa) and sintering at 1200 °C for shorter time of 10 min, nanoceramic with density of 92% and grain size smaller than 100 nm was produced. Apatite was the main crystalline phase in sintered ceramic. It presented minor impurity phase of La_2SiO_5 as a secondary phase.

To prepare a mixed composite electrolyte of the apatite-type lanthanum silicates $La_{10}Si_6O_{27}$ (LSO) and $Sm_{0.2}Ce_{0.8}O_{1.9}$ (SDC), Shi Qingle *et al.* [57] used the sol-gel process to synthetize LSO. TEOS was added to a mixture solution of $La(NO_3)_3 \cdot 6H_2O$, distilled water and acidic acid to result a purple clear sol. Clear gel was obtained by refluxing for 1–2 h at 80 °C. After drying and calcing the gel at 1000 °C, the apatite-type LSO phase was obtained, confirmed by XRD patterns. No secondary phase was observed.

El Nahrawy and co-workers [58] synthesized nanostructured pure silica and oxyapatite lanthanum silicate ($10La_2O_3$-$90SiO_2$) nanoparticles using sol–gel technique. Lanthanum nitrate was added in sol–gel derived silica matrix. Pure SiO_2 was prepared under vigorous stirring (950 rpm) for 2h from TEOS dissolved in excess of absolute ethanol and distilled water in the presence of HNO_3. The lanthanum silicate system was obtained from TEOS and mixed solution of lanthanum nitrate, distilled water, ethanol and nitric acid by vigorous stirring for 2h at 75 °C. After overnight, a clear gel resulted. Thermal treatment of the gel was performed at temperatures from 200 to 800 °C, to study the effects of H_2O:TEOS ratio on the structural and spectroscopic properties of La_2O_3-SiO_2 system, and the effect of La_2O_3 content on the SiO_2. They concluded that the crystallinity of $10La_2O_3$-$90SiO_2$ powders increases with increasing calcination temperature. The size of nanoparticles mainly ranged from 10 to 27 nm, depends on H_2O content. XRD and FTIR studies confirmed the homogeneity of the constituents La_2O_3 and SiO_2 in lanthanum silicate powders. The optical properties changed gradually with increasing the calcination temperature. The prepared nanoparticles samples presented spherical morphology.

A series of $La_{10-x}Si_6O_{27-1.5x}$ precursor powders were produced *via* a modified sol-gel process [59]. Stoichiometric ratios of $La(NO_3)_3 \cdot 6H_2O$ and TEOS were mixed followed by addition of citric acid and ethylene glycol, to form a transparent solution. Then it was heated in water bath at 80 °C for about 3~4h to result a sol with light yellow color. The sol was dried at 90 °C in an oven, ground, calcined at 600 °C for 4h. Pure apatite-type precursor powders were attained by calcining at 1000°C for 4h. It was concluded the increase of La deficiency resulted in lattice distortion and more cation vacancy at the position of 4f, making La^{3+} to diffuse more easily. La^{3+} accumulated at the grain boundaries, inhibiting the conduction of oxide ions, thus reducing the conductivity. The highest conductivity was 0.03 S/cm at 800 °C for $La_{10}Si_6O_{27}$ sintered at 1600 °C for 6h.

Coprecipitation Synthesis Method

The oldest technique to obtain homogeneous mixed oxides is the coprecipitation synthesis method. It consists in precipitating, usually, hydrous or oxalates of the in aqueous media, where the precursors salts are homogenously mixed. Usually the separation of the precipitated and the liquid solution is made by filtration process. The homogeneous oxide mixture is achieved when the precipitate after washing is thermally treated. The oxide mixture is in turn thermal treated again to attain the desired crystalline phase. Control of the process parameters is very important to ensure the homogeneity oxide mixture to attain the correct composition, appropriate morphology and purity of the final product. Temperature, pH, the reagents concentration and mixing rates are some of those parameters. In an ideal coprecipitation synthesis method, the precipitation of all components is desired to occur together at the same time. Actually, inhomogeneity at microscopic level may be generated if some individual compound precipitates at different time. In the coprecipitation technique, agglomerates are generally produced during the calcination process. However, by optimizing and control the parameters of the method, it produces stoichiometric powders of high purity and fine particle size at a relatively moderate cost. Therefore, it is useful to applying in industrial ceramic powders production.

Yang *et al.* [60] used a low cost coprecipitation method to synthesize LSA precursor powders. La_2O_3 dissolved in 1M HNO_3 and TEOS dissolved in ethanol were raw materials. They synthesized $La_{9.33+r}Si_6O_{26+1.5x}$ powders by coprecipitation method with ammonia as precipitant. Ammonia (~28 volume %) with a certain amount of polyethylene glycol (PEG200) was diluted with the mixture of de-ionized water and ethanol. The pH value of this mixture of diluted ammonia was adjusted to be 9–10. The solution containing metal ions was added into diluted ammonia by drop wise with vigorous stirring. The obtained precipitate was separated by centrifugation, washed with ethanol, dried at 80 °C and ground. Well-dispersed nanopowders (*ca.* 70 nm) with pure hexagonal LSA phase were obtained by calcining at 900 °C for 9h. Dense LSA ceramic with relative density of 98% could be achieved at 1550 °C.

Highly sinterable $La_{10}Si_6O_{27}$ and $La_{10}Si_{5.5}M_{0.5}O_{27}$ (M = Mg, and Al) nanopowders were synthesized by homogeneous coprecipitation method. $La(NO_3)_3.6H_2O$, $Mg(NO_3)_2.6H_2O$ $Al(NO_3)_3.9H_2O$ and TEOS were used as starting materials and diethylamine (DEA) as a precipitant [61]. By this simple and effective homogeneous coprecipitation procedure, weakly agglomerated nanopowders (~30 nm) with high crystallinity single apatite-phase were achieved. High dense ceramic (>95%) resulted from sintering the powder at 1400 °C.

Kharlamova *et al.* [62] presented a study of peculiarities of genesis of the structure and texture of apatite-type lanthanum silicates. Both undoped and $La_{10-y+x/3}(SiO_4)_{6-x}(AlO_4)_xO_{3-3y/2}$ (x=0.25; 0.5; 1; y=0.33; 0.67) samples were prepared by co- precipitation, at early stages of synthesis. They concluded that an amorphous lanthanum silicate with the short-range ordering typical for ortho-diorthosilicate was shown to be formed in the course of precipitation, while its crystallization proceeds *via* rearrangement of the polymeric structure of primary particles during subsequent thermal treatment. Various physic-chemical methods: XRD, IR spectroscopy, thermal analysis, low-temperature adsorption of nitrogen, 27Al and 29Si MAS NMR spectroscopy, and TEM with EDX analysis were used to explain the precipitation process.

Pechini Synthesis Method

The Pechini method was proposed in 1967 [63] to use for capacitors component production. The technique was suitable for the deposition of dielectric film of titanates and niobates of lead for use in capacitors.

Currently, Pechini method is well known as method for preparing metal oxide powders [64]. The method consists on polymerization of metal citrates by ethylene glycol. In general, citric, tartaric and glycolic acids with aqueous solution of metals are used to form metal polybasic chelates. By further heat-treatment, a viscous resin, followed by a rigid transparent, glassy gel and a fine dispersed oxide powder is performed. The process was adapted for laboratory synthesis of multicomponent oxide materials. Pechini method [64] allows preparing complex compositions with good homogeneity due to the molecular level mixing of precursors in the solution. The low temperature for resin decomposition and its simplicity are advantages of the method. The disadvantage is the use of toxic ethylene glycol and substantial volumes of organic reagents generated.

Kioupis *et al.* [65] reported modified Pechini method. They studied the effect of the stoichiometry on the structure and morphology of intermediate precursor of $La_{9.83-x}Sr_xSi_6O_{26+\delta}$ (0≤x≤0.50) as well as on the final product. La_2O_3 $SrCO_3$ and SiO_2 sol in ethylene glycol (30% *w/w*) were raw materials. Citric acid was dissolved in deionized water to prepare resins that was mixed to ethylene glycol with adjusting molar ratio to 3, resulting clear solution. Then, stoichiometric nitric cation (La, Sr and Si) was added to the above clear solution, with continuous stirring at 60 °C for at least 30 min. The resulted clear mixed solution was 15 min thermal treated in an adapted domestic microwave (MW) oven in order to remove excess of water and to assist the polyesterification reaction. Brown resin has formed by heating the product of polyesterification reaction, at 150 °C. Pure $La_{9.83}Si_6O_{26.75}$, $La_{9.38}Sr_{0.45}$ $Si_6O_{26.52}$ and $La_{9.33}Sr_{0.50}Si_6O_{26.50}$ were prepared from this resin calcined at temperatures between 600 and 1400 °C for period of 3 to 20 h and subsequent 20h sintering at 1400 °C. In final products of $La_{9.68}Sr_{0.15}Si_6O_{26+\delta}$ and $La_{9.53}Sr_{0.30}Si_6O_{26+\delta}$ compounds, minor traces (<3.5%) of $La_2Si_2O_7$ secondary phase was observed.

Vanmeensel *et al.* [66] prepared nano-sized powders with pure crystalline phase of Al doped lanthanum silicate, $La_{9.83}Al_{1.5}Si_{4.5}O_{26}$, and Fe doped lanthanum, $La_{9.83}Fe_{1.5}Si_{4.5}O_{26}$, by sol–gel Pechini processing by calcining the gel at 900 °C for 8 h.

Hydrothermal Synthesis Method

Hydrothermal synthesis consists in submitting the aqueous solution of precursors in a vessel, at high temperature (>100°C) and pressure higher than a few atmospheres [67]. The environment of high temperature and pressure can reduce free energies for stabilizing the equilibrium of phases that not be attained at normal conditions [68]. In the process, dissolution and precipitation or in *in-situ* transformation or both occur to form ceramic oxide particles. For example, particles of the product may precipitated out from oxides and hydroxides of component dissolved into the supersaturated solution, due to the high temperature and pressure.

Sometimes, the process needs to add mineralizer such as bases, if the suspended solids not solubilize in aqueous solution. In some cases, the particles of the product form by polymorphic or chemical phase transformation [69], *via* another in-situ transformation mechanism. Parameters such as initial pH, pressure, temperature and time of synthesis have high influence to the kinetics of the process.

The synthesis of undoped $La_{9.33}Si_6O_{26}$ and doped apatites ($La_9CaSi_6O_{26.5}$, $La_9SrSi_6O_{26.5}$, and $La_9BaSi_6O_{26.5}$) using a hydrothermal method is reported by Noviyanti and co-workers [70]. From basic solution of La_2O_3, Na_2SiO_4, $BaCO_3$, $CaCO_3$, and $SrCO_3$, heated at 240 °C in an autoclave for 3 days, apatite-type phases of La9.33Si6O26 and Sr and Ba doped apatites have prepared successfully. Although the apatite phase has formed in 3 days, temperatures higher than 1600 °C was required to obtain dense ceramics, while sintering at lower temperatures (1100 °C), even for a longer time (17 h), does not resulted in a dense apatite ceramic.

In the hydrothermal synthesis performed by Kitamura *et al.* [71], stoichiometric amounts of $LaCl_3.6H_2O$ and SiO_2 were mixed in 4.2 mol. L^{-1} NaOH solution for 1h and the mixture was transferred into autoclaves. Synthesis was carried out in period of 24-168h at temperatures of 180-230 °C to obtain a precipitate. Subsequently, the precipitate was water washed and dried in period of one day at 100 °C, resulted the final powder product. Samples synthesized at higher temperature than 200 °C, secondary phases take place. Instead, the samples synthesized at 180 °C presented single phase. Particle size of the powder was between 200-300nm.

Freeze-Drying Synthesis

In freeze-drying [72] synthesis, the solvent contained in small droplet in which are the solutions of cations precursors, is slowly sublimed. This method allows an excellent control over composition, homogeneity and impurity levels. The rapid vaporization or slow sublimation of the solvent reduces the problematic agglomeration of the particles related to the large surface tension of vapor-liquid interface.

Chesnaud *et al.* [28], have synthesized a series of polycrystalline lanthanum silicates, $(La_{9.33+x}(SiO_4)_6O_{2+3x/2}$ $(0 \leq x \leq 0.67))$, using an optimized freeze drying method, starting with precursor cation solution modified by acetic acid. Initially, mixed solution of lanthanum acetate, TEOS and acetic acid was conditioned at pH 3.8. After attaining the pH, it is essential the solution be clear to ensure the stoichiometric homogeneity at the nanometric scale of the precursor cations. Then, this clear conditioned mixture solution is adding, by spraying, into the liquid nitrogen to form frozen droplets. The freeze-dried product calcined at 900 °C, produced an ultrafine and very homogeneous nanopowders within in size range from 100 to 200 nm. Those particles were submitted to conventional and to spark plasma sintering (SPS). Dense ceramic, at relatively low temperatures, were prepared.

Plasma Spraying Synthesis

An advanced method of synthesis is the plasma spraying [73] technique. It consists on injecting high temperature melted raw materials into plasma flame environment. This process is usual to produce protective coating to enhancing its performance.

Complexity of the process is a variety of many requirements of plasma flame. The temperature, velocity and the enthalpy density must be high. Adding to those factors, it requires an active environment and both heating and cooling rates extremely high. Those requirements are advantageous to its use to many uncommon applications, including particle spheroidization, in some reaction of synthesis, process that a rapid solidification is required, *etc.* The complexity and relative high cost are disadvantages of the method.

Apatite-type oxide powders of $La_{10}(SiO_4)_6O_3$ have been elaborated by Gao *et al.* [74] through atmospheric plasma spraying (APS) using micro-scale mixtures of La_2O_3 and SiO_2 as starting materials. The dried starting materials were mixed with appropriate amounts of polyvinyl alcohol (PVA), heated, crushed and screened. These powders were melted or partially melted by heating of the plasma jet with extremely high temperature (>10000K) and they were collected after rapid solidification. This process has successfully synthesized lanthanum silicate $La_{10}Si_6O_{27}$ powders. The formation process of $La_{10}Si_6O_{27}$ was significantly predigested compared with those prepared by solid-state reaction or sol-gel process. The starting particles size of starting materials has a great influence on the formation of $La_{10}Si_6O_{27}$ particles.

Chesnaud *et al.* [28] have prepared lanthanum silicates with general formula, $La_{10-x}(SiO_4)_6O_{3-1.5x}$, by spray plasma synthesis (SPS). In the method the sintering temperature and time were reduced. Under 100 MPa pressure and temperature of 1473 and 1773 K, a dense transparent ceramic has produced. At lower pressure (75 Mpa) and temperature at 1673 K Porras-Va´zquez *et al.* [75] also obtained fully dense, ~99%, optical transparent ceramics of $La_{9.33}(SiO_4)_6O_2$.

High-Energy Ball Milling Synthesis

In 1960, Benjamin *et al.* [76] developed the high-energy ball milling technique at a nickel company. In the process, also called mechanical alloying, a mixture of powders was submitted to high-energy collision from the balls. The process produced fine Al_2O_3, Y_2O_3 and ThO_2 nickel based super-alloys. By conventional powder metallurgy methods, those products are not possible to make. Their innovation has changed the traditional method of producing materials by high temperature synthesis. High-energy ball milling process may changing the conditions in which chemical reaction can occur, *i.e.*, by varying the reactivity of as-milled powders or by inducing chemical reactions during milling (mechanochemistry).

Rodríguez-Reyna *et al.* [77 - 79] have used high-energy dry ball milling in a planetary ball to obtain lanthanum silicates. La_2O_3 and amorphous or low cristobalite silica were dry-milled with zirconia balls in air using a planetary ball mill. The speed of rotating disc was 350 rpm, reversed every 20 min. From observation of the powder milled XRD patterns, and IR and Raman spectra, they revealed that in both cases, starting with amorphous silica or low cristobalite SiO_2, this mechanical procedure not induces chemical reaction to form an amorphous precursor. Nevertheless, the use of amorphous silica is favorable, according to the investigation from XRD data of activation energies for crystallite growth.

Kobayashi *et al.* [47] produced high sinterable lanthanum oxyapatite powders *via* high-energy planetary ball-milling followed solid-sate reaction.

Combined Synthesis Methods

Recently, Macedo *et al.* [78] proposed an innovative chemical route, which combines coprecipitation, and sol-gel approaches to synthesize LSA. Apatite powders with nominal composition $La_{10}Si_6O_{27}$ were synthesized by the proposed method. The correlation between sintering temperature and electrical properties of ceramics was reported. $La(NO_3)_3.6H_2O$ and TEOS were as raw materials. Mixed clear solution of TEOS dissolved into n-propanol and aqueous La nitrate was dropped into a NH_4OH solution (28 wt.%) under vigorous and constant stirring until a final pH of 10. The stirring process was keeping for 3 h at room temperature to a white gel formation, endorsing the polymerization. The resulted gel, which was separated by filtration, then, it was washed with n-propanol, followed by drying overnight at 100 °C and calcined between 500 and 900 °C. Rietveld refinement of the XRD data showed the synthesized powder particle size is lower than 100 nm, and SEM micrographs revealed high level of agglomeration.

A new sol–gel process water-based was proposed by Yamagata *et al.* [79] to synthetize apatite-type lanthanum silicate. The process combines sol-gel and precipitation techniques. Firstly, spherical shaped gel of silica was prepared from Na_2SiO_3 solution by HCl catalyzed reaction, in presence of a surfactant to attain aerogel characteristics. Obtained aerogel silica was water washed and added to a stoichiometric (to give $La_{9.56}(SiO_4)_6O_{2.33}$) amount of La nitric solution to lanthanum embedding on the gel. Subsequently the pH of the gel La embedded was adjusted to 11 by adding of NH_4OH solution. At this pH, lanthanum hydroxide precipitates resulting gel of silica together lanthanum hydroxide, which was separated by filtration, washed with water and ethanol. Overnight thermal treatment at 80 °C and calcining 900 °C for 4h, single crystalline phase of apatite powder was achieved. Particle size, evaluated by specific surface area (27.3 m^2 g^{-1} and 34.2 m^2 g^{-1}) is between 30 and 40 nm.

Misso *et al.* [80] reported a modified sol-gel process. They combined sol-gel process with molten salt approaches to synthesize LSA, with the composition $La_{9.33}Si_6O_{26}$. Lanthanum oxide solubilized in 15 M nitric acid. This lanthanum nitric solution was added to stoichiometric amounts of sodium silicate solution, in such a way as the pH of the mixture be enough acidic to occurs the acid catalysis reaction to the gelatinization of silica. The gel was calcined at 900 °C for 1h to allow lanthanum salt melting. The white obtained powder was washed with water, filtered and dried overnight at 80 °C. Subsequently, it was calcined at 900 °C to crystallize apatite phase. Following the same procedure of synthesis proposed by Misso *et al.*, a high specific surface area powder (32.70 m^2. g^{-1}) was obtained by Elias [81]. XRD patterns of the precursor of $La_{9.33}Si_6O_{26}$ calcined at 900 °C for 1h is shown in Fig. (**4**).

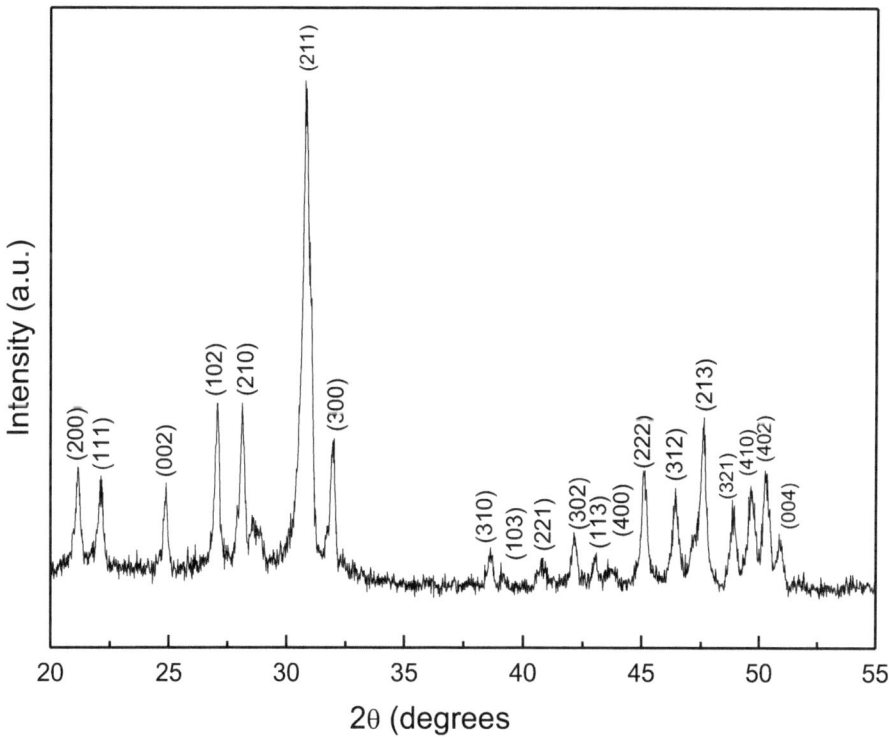

Fig. (4). XRD of the precursor powder of $La_{9.33}Si_6O_{26}$ calcined at 900 °C for 1h.

From XRD patterns in Fig. (**4**), apatite is the main crystalline phase by comparison with JCPDS 49-0443 data. This powder pressed to pellet and sintered for 4h at 1200 °C, pure and high crystalline apatite phase is verified as showed in Fig. (**5**).

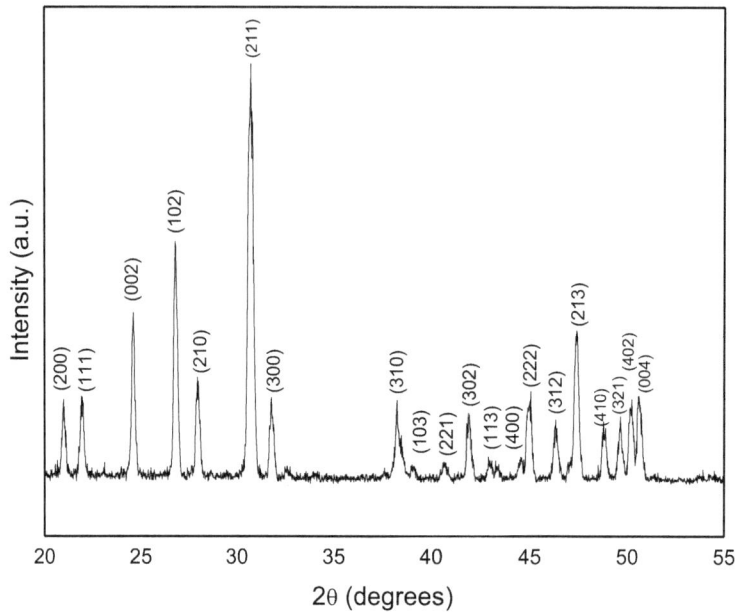

Fig. (5). XRD of the ceramic body sintered at 1200 °C for 4h, obtained from precursor of $La_{9.33}Si_6O_{26}$ calcined at 900 °C for 1h.

The ceramic sample, sintered at 1300 °C for 4h, presented the identical XRD patterns of $La_{9.33}Si_6O_{26}$ pure apatite as observed in Fig. (**5**). SEM micrographs of ceramics sintered at 1200 °C and 1300 °C for 4h are presented in (Fig. **6a** and **6b**), respectively.

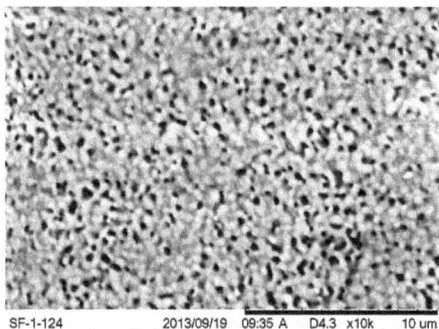 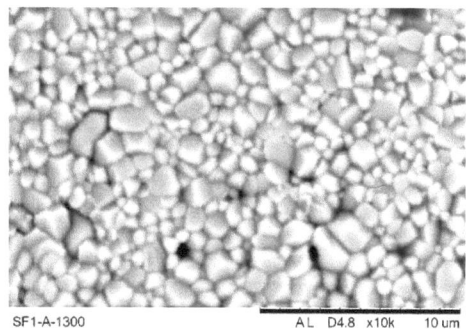

(a)   (b)

Fig. (6). SEM micrographs of the ceramic sintered at 1200 °C for 4h (**a**) and it sintered at 1300 °C for 4h (**b**).

In Fig. (**6a**), a porous microstructure is observed, while in Fig. (**6b**) a highly dense ceramic with well-defined grains boundaries are present, and the porosity is lowest. From that observation, it can be conclude that temperature of 1200 °C is adequate to crystallize the apatite phase (Fig. **5**) but high porosity remain in ceramic body (Fig. **6a**). The relative densities obtained by Elias [81] clear confirm it. Ceramic sintered at 1200 ^{0}C showed 79.5% of relative density wereas the sample sintered at 1300 °C, 93.3% theoretical density was achieved, that could be adequate for its application in IT-SOFC [18]. The sintering temperature of 1300 °C used in the work of Elias [81] is about 400 °C lower than that of the conventional usual temperature [21]. Tao *et al.* [31] in their study, sintered at 1400 °C in periods of 20h and three-day, but the relative attained densities were 69% and 74%, respectively. Even increasing the temperature and time to 1500 °C and 22h, the resulted ceramic was 80% theoretical density, which is not suitable for its IT-SOFC electrolyte application request. In the Table **2** relative densities of $La_{9.33}Si_6O_{26}$ ceramic body obtained by some investigation are presented.

By comparison of Table **2** data, the sol-gel water-based combined with molten salt technique proposed by Elias [81] is promising method to synthesize apatite-type lanthanum silicate powder.

Table 2. Relative densities of $La_{9.33}Si_6O_{26}$ ceramic body.

Method	Sintering Condition Temperature (°C), Time (h)	Relative Density (%)	Reference
Proposed by Elias	1300, 4	93.3	[81]
Conventional solid-state	1450, 96	~95	[75]
Conventional solid state	1700, -	~95	[21]
Sol gel	1500, 10	90	[82]
Molten salt	1500, 4	>90	[83]
Sol gel	1450, 20	92	[84]

CONFLICT OF INTEREST

The authors confirm that they have no conflict of interest to declare for this publication.

ACKNOWLEDGEMENTS

Authors acknowledge financial support from the CNPq and FAPESP.

REFERENCES

[1]   Singhal, S.C.; Kendall, K., Eds. *High-temperature solid oxide fuel cells: fundamentals, design and applications*; Elsevier, **2003**.

[2]   Fergus, J.; Hui, R.; Li, X.; Wilkinson, D.P.; Zhang, J., Eds. *Solid oxide fuel cells: materials properties and performance*; CRC press, **2008**.

[3]   Sun, C.; Hui, R.; Roller, J. Cathode materials for solid oxide fuel cells: a review. *J. Solid State Electrochem.,* **2009**, *14*, 1125-1144.
[http://dx.doi.org/10.1007/s10008-009-0932-0]

[4]   Kendrick, E.; Islam, M.S.; Slater, P.R. Developing apatites for solid oxide fuel cells: insight into structural, transport and doping properties. *J. Mater. Chem.,* **2007**, *17*, 3104-3111.
[http://dx.doi.org/10.1039/b704426g]

[5]   Nakayama, S.; Sakamoto, M. Electrical properties of new type high oxide ionic conductor $RE_{10}Si_6O_{27}$ (RE= La, Pr, Nd, Sm, Gd, Dy). *J. Eur. Ceram. Soc.,* **1998**, *18*, 1413-1418.
[http://dx.doi.org/10.1016/S0955-2219(98)00032-6]

[6]   Nakayama, S.; Sakamoto, M.; Higuchi, M.; Kodaira, K.; Sato, M.; Kakita, S.; Suzuki, T.; Itoh, K. Oxide ionic conductivity of apatite type $Nd_{9.33}(SiO_4)_6O_2$ single crystal. *J. Eur. Ceram. Soc.,* **1999**, *19*, 507-510.
[http://dx.doi.org/10.1016/S0955-2219(98)00215-5]

[7]   Nakayama, S.; Sakamoto, M. Ionic conductivities of apatite-type $La_x(GeO_4)_6O_{1.5x-12}$ (x= 8–9.33) polycrystals. *J. Mater. Sci. Lett.,* **2001**, *20*, 1627-1629.
[http://dx.doi.org/10.1023/A:1017914229923]

[8]   Lambert, S.; Vincent, A.; Bruneton, E.; Beaudet-Savignat, S.; Guillet, F.; Minot, B.; Bouree, F. Structural investigation of $La_{9.33}Si_6O_{26}$- and $La_9AESi_6O_{26+\delta}$-doped apatites-type lanthanum silicate (AE= Ba, Sr and Ca) by neutron powder diffraction. *J. Solid State Chem.,* **2006**, *179*, 2602-2608.
[http://dx.doi.org/10.1016/j.jssc.2006.04.056]

[9]   Tolchard, J.R.; Sansom, J.E.; Islam, M.S.; Slater, P.R. Structural studies of apatite-type oxide ion conductors doped with cobalt. *Dalton Trans.,* **2005**, *7*(7), 1273-1280.
[http://dx.doi.org/10.1039/b418992b] [PMID: 15782264]

[10]   Ali, R.; Yashima, M.; Matsushita, Y.; Yoshioka, H.; Izumi, F. Crystal structure and electron density in the apatite-type ionic conductor $La_{9.71}(Si_{5.81}Mg_{0.18})O_{26.37}$. *J. Solid State Chem.,* **2009**, *182*, 2846-2851.
[http://dx.doi.org/10.1016/j.jssc.2009.07.053]

[11]   Kendrick, E.; Sansom, J.E.; Tolchard, J.R.; Islam, M.S.; Slater, P.R. Neutron diffraction and atomistic simulation studies of Mg doped apatite-type oxide ion conductors. *Faraday Discuss.,* **2007**, *134*, 181-194.
[http://dx.doi.org/10.1039/B602258H] [PMID: 17326569]

[12]   Slater, P.R. An apatite for fast oxide ion conduction. *Chem. Commun,* **2003**, 1486-1487.
[http://dx.doi.org/10.1039/B301179H]

[13]   Kendrick, E.; Islam, M.S.; Slater, P.R. Atomic-scale mechanistic features of oxide ion conduction in apatite-type germanates. *Chem. Commun. (Camb.),* **2008**, *6*(6), 715-717.
[http://dx.doi.org/10.1039/B716814D] [PMID: 18478700]

[14]   León-Reina, L.; Porras-Vázquez, J.M.; Losilla, E.R.; Aranda, M.A. Phase transition and mixed oxide-proton conductivity in germanium oxy-apatites. *J. Solid State Chem.,* **2007**, *180*, 1250-1258.
[http://dx.doi.org/10.1016/j.jssc.2007.01.023]

[15]   León-Reina, L.; Porras-Vázquez, J.M.; Losilla, E.R.; Sheptyakov, D.V.; Llobet, A.; Aranda, M.A. Low temperature crystal structures of apatite oxygen-conductors containing interstitial oxygen. *Dalton Trans.,* **2007**, *20*(20), 2058-2064.
[http://dx.doi.org/10.1039/B616211H] [PMID: 17502939]

[16]   León-Reina, L.; Losilla, E.R.; Martínez-Lara, M.; Bruque, S.; Llobet, A.; Sheptyakov, D.V.; Aranda, M.A. Interstitial oxygen in oxygen-stoichiometric apatites. *J. Mater. Chem.*, **2005**, *15*, 2489-2498.
[http://dx.doi.org/10.1039/b503374h]

[17]   León-Reina, L.; Losilla, E.R.; Martínez-Lara, M.; Martín-Sedeño, M.C.; Bruque, S.; Núnez, P.; Sheptyakov, D.V.; Aranda, M.A. High oxide ion conductivity in Al-doped germanium oxyapatite. *Chem. Mater.*, **2005**, *17*, 596-600.
[http://dx.doi.org/10.1021/cm048361r]

[18]   León-Reina, L.; Losilla, E.R.; Martínez-Lara, M.; Bruque, S.; Aranda, M.A. Interstitial oxygen conduction in lanthanum oxy-apatite electrolytes. *J. Mater. Chem.*, **2004**, *14*, 1142-1149.
[http://dx.doi.org/10.1039/B315257J]

[19]   Yoshioka, H. Oxide ionic conductivity of apatite-type lanthanum silicates. *J. Alloys Compd.*, **2006**, *408*, 649-652.
[http://dx.doi.org/10.1016/j.jallcom.2004.12.180]

[20]   Yoshioka, H.; Nojiri, Y.; Tanase, S. Ionic conductivity and fuel cell properties of apatite-type lanthanum silicates doped with Mg and containing excess oxide ions. *Solid State Ion.*, **2008**, *179*, 2165-2169.
[http://dx.doi.org/10.1016/j.ssi.2008.07.022]

[21]   Abram, E.J.; Sinclair, D.C.; West, A.R. A novel enhancement of ionic conductivity in the cation-deficient apatite $La_{9.33}(SiO_4)_6O_2$. *J. Mater. Chem.*, **2001**, *11*, 1978-1979.
[http://dx.doi.org/10.1039/b104006p]

[22]   Vincent, A.; Savignat, S.B.; Gervais, F. Elaboration and ionic conduction of apatite-type lanthanum silicates doped with Ba, $La_{10-x}Ba_x(SiO_4)_6O_{3-x/2}$ with x= 0.25–2. *J. Eur. Ceram. Soc.*, **2007**, *27*, 1187-1192.
[http://dx.doi.org/10.1016/j.jeurceramsoc.2006.05.090]

[23]   Tao, S.; Irvine, J.T. Preparation and characterisation of apatite-type lanthanum silicates by a sol-gel process. *Mater. Res. Bull.*, **2001**, *36*, 1245-1258.
[http://dx.doi.org/10.1016/S0025-5408(01)00625-0]

[24]   Sansom, J.E.; Richings, D.; Slater, P.R. A powder neutron diffraction study of the oxide-io--conducting apatite-type phases, $La_{9.33}Si_6O_{26}$ and $La_8Sr_2Si_6O_{26}$. *Solid State Ion.*, **2001**, *139*, 205-210.
[http://dx.doi.org/10.1016/S0167-2738(00)00835-3]

[25]   Sansom, J.E.; Tolchard, J.R.; Slater, P.R.; Islam, M.S. Synthesis and structural characterisation of the apatite-type phases $La_{10-x}Si_6O_{26+z}$ doped with Ga. *Solid State Ion.*, **2004**, *167*, 17-22.
[http://dx.doi.org/10.1016/j.ssi.2003.12.014]

[26]   Shaula, A.L.; Kharton, V.V.; Marques, F.M. Oxygen ionic and electronic transport in apatite-type $La_{10-x}(Si,Al)_6O_{26\pm\delta}$. *J. Solid State Chem.*, **2005**, *178*, 2050-2061.
[http://dx.doi.org/10.1016/j.jssc.2005.04.018]

[27]   Panteix, P.J.; Julien, I.; Bernache-Assollant, D.; Abelard, P. Synthesis and characterization of oxide ions conductors with the apatite structure for intermediate temperature SOFC. *Mater. Chem. Phys.*, **2006**, *95*, 313-320.
[http://dx.doi.org/10.1016/j.matchemphys.2005.06.040]

[28]   Chesnaud, A.; Dezanneau, G.; Estournès, C.; Bogicevic, C.; Karolak, F.; Geiger, S.; Geneste, G. Influence of synthesis route and composition on electrical properties of $La_{9.33+x}Si_6O_{26+3x/2}$ oxy-apatite compounds. *Solid State Ion.*, **2008**, *179*, 1929-1939.
[http://dx.doi.org/10.1016/j.ssi.2008.04.035]

[29] Chesnaud, A.; Bogicevic, C.; Karolak, F.; Estournès, C.; Dezanneau, G. Preparation of transparent oxyapatite ceramics by combined use of freeze-drying and spark-plasma sintering. *Chem. Commun. (Camb.),* **2007**, *15*(15), 1550-1552.
[http://dx.doi.org/10.1039/B615524C] [PMID: 17406704]

[30] Kendrick, E.; Islam, M.S.; Slater, P.R. Atomic-scale mechanistic features of oxide ion conduction in apatite-type germanates. *Chem. Commun. (Camb.),* **2008**, *6*(6), 715-717.
[http://dx.doi.org/10.1039/B716814D] [PMID: 18478700]

[31] Tao, S.; Irvine, J.T. Preparation and characterisation of apatite-type lanthanum silicates by a sol-gel process. *Mater. Res. Bull.,* **2001**, *36*, 1245-1258.
[http://dx.doi.org/10.1016/S0025-5408(01)00625-0]

[32] Kendrick, E.; Headspith, D.; Orera, A.; Apperley, D.C.; Smith, R.I.; Francesconi, M.G.; Slater, P.R. An investigation of the high temperature reaction between the apatite oxide ion conductor $La_{9.33}Si_6O_{26}$ and NH_3. *J. Mater. Chem.,* **2009**, *19*, 749-754.
[http://dx.doi.org/10.1039/B808215D]

[33] Imaizumi, K.; Toyoura, K.; Nakamura, A.; Matsunaga, K. Strong correlation in 1D oxygen-ion conduction of apatite-type lanthanum silicate. *J. Phys. Condens. Matter,* **2015**, *27*(36), 365601.
[http://dx.doi.org/10.1088/0953-8984/27/36/365601] [PMID: 26302221]

[34] Tolchard, J.R.; Islam, M.S.; Slater, P.R. Defect chemistry and oxygen ion migration in the apatite-type materials $La_{9.33}Si_6O_{26}$ and $La_8Sr_2Si_6O_{26}$. *J. Mater. Chem.,* **2003**, *13*, 1956-1961.
[http://dx.doi.org/10.1039/b302748c]

[35] Miura, H. CellCalc: A Unit Cell Parameter Refinement Program on Windows Computer. *J Cryst Soc Jpn,* **2003**, *45*, 145-147.
[http://dx.doi.org/10.5940/jcrsj.45.145]

[36] Sansom, J.E.; Kendrick, E.; Tolchard, J.R.; Islam, M.S.; Slater, P.R. A comparison of the effect of rare earth *vs* Si site doping on the conductivities of apatite-type rare earth silicates. *J. Solid State Electrochem.,* **2006**, *10*, 562-568.
[http://dx.doi.org/10.1007/s10008-006-0129-8]

[37] Béchade, E.; Julien, I.; Iwata, T.; Masson, O.; Thomas, P.; Champion, E.; Fukuda, K. Synthesis of lanthanum silicate oxyapatite materials as a solid oxide fuel cell electrolyte. *J. Eur. Ceram. Soc.,* **2008**, *28*, 2717-2724.
[http://dx.doi.org/10.1016/j.jeurceramsoc.2008.03.045]

[38] Kendrick, E.; Islam, M.S.; Slater, P.R. Developing apatites for solid oxide fuel cells: insight into structural, transport and doping properties. *J. Mater. Chem.,* **2007**, *17*, 3104-3111.
[http://dx.doi.org/10.1039/b704426g]

[39] McFarlane, J.; Barth, S.; Swaffer, M.; Sansom, J.E.; Slater, P.R. Synthesis and conductivities of the apatite-type systems, $La_{9.33+x}Si_{6-y}M_yO_{26+z}$ (M= Co, Fe, Mn) and $La_8Mn_2Si_6O_{26}$. *Ionics,* **2002**, *8*, 149-154.
[http://dx.doi.org/10.1007/BF02377766]

[40] Fukuda, K.; Asaka, T.; Okino, M.; Berghout, A.; Béchade, E.; Masson, O.; Julien, I.; Thomas, P. Anisotropy of oxide-ion conduction in apatite-type lanthanum silicate. *Solid State Ion.,* **2012**, *217*, 40-45.
[http://dx.doi.org/10.1016/j.ssi.2012.04.018]

[41] Panteix, P.J.; Julien, I.; Bernache-Assollant, D.; Abelard, P. Synthesis and characterisation of oxide ions conductors with the apatite structure for intermediate temperature SOFC. *Key Eng. Mater.,* **2004**, *264*, 1177-1180.
[http://dx.doi.org/10.4028/www.scientific.net/KEM.264-268.1177]

[42]   Lee, D.J.; Lee, S.G.; Noh, H.J.; Jo, Y.W. Characteristics of Lanthanum Silicates Electrolyte for Solid Oxide Fuel Cells. *Trans Electr Electron Mater,* **2015**, *16*, 194-197.
[http://dx.doi.org/10.4313/TEEM.2015.16.4.194]

[43]   Cao, X.G.; Jiang, S.P. Synthesis and characterization of lanthanum silicate oxyapatites co-doped with A (A= Ba, Sr, and Ca) and Fe for solid oxide fuel cells. *J. Mater. Chem. A Mater. Energy Sustain.,* **2014**, *2*, 20739-20747.
[http://dx.doi.org/10.1039/C4TA04616A]

[44]   Guang-Chao, Y.; Hong, Y.; Lin-Hong, Z.; Mei-Ling, S.; Jun-Kai, Z.; Xiao-Jun, X.; Ri-Dong, C.; Xin, W.; Wei, G.; Qi-Liang, C. Crystal structure and ionic conductivity of Mg-doped apatite-type lanthanum silicates $La_{10}Si_{6-x}Mg_xO_{27-x}$ (x= 0–0.4). *Chin. Phys. B,* **2014**, *23*, 048202.
[http://dx.doi.org/10.1088/1674-1056/23/4/048202]

[45]   Xiang, J.; Liu, Z.G.; Ouyang, J.H.; Yan, F.Y. Influence of sintering parameters on microstructure and electrical conductivity of $La_{10}Si_6O_{27}$ ceramics. *Ceram. Int.,* **2014**, *40*, 2401-2410.
[http://dx.doi.org/10.1016/j.ceramint.2013.08.012]

[46]   Kobayashi, K.; Hirai, K.; Suzuki, T.S.; Uchikoshi, T.; Akashi, T.; Sakka, Y. Sinterable powder fabrication of lanthanum silicate oxyapatite based on solid-state reaction method. *J. Ceram. Soc. Jpn.,* **2015**, *123*, 274-279.
[http://dx.doi.org/10.2109/jcersj2.123.274]

[47]   Pons, A.; Jouin, J.; Béchade, E.; Julien, I.; Masson, O.; Geffroy, P.M.; Mayet, R.; Thomas, P.; Fukuda, K.; Kagomiya, I. Study of the formation of the apatite-type phases $La_{9.33+x}(SiO_4)_6O_{2+3x/2}$ synthesized from a lanthanum oxycarbonate $La_2O_2CO_3$. *Solid State Sci.,* **2014**, *38*, 150-155.
[http://dx.doi.org/10.1016/j.solidstatesciences.2014.10.013]

[48]   Shao, Z.; Zhou, W.; Zhu, Z. Advanced synthesis of materials for intermediate-temperature solid oxide fuel cells. *Prog. Mater. Sci.,* **2012**, *57*, 804-874.
[http://dx.doi.org/10.1016/j.pmatsci.2011.08.002]

[49]   Lalena, JN; Cleary, DA; Carpenter, E; Dean, NF *Inorganic materials synthesis and fabrication*; John Wiley & Sons: USA, **2008**.
[http://dx.doi.org/10.1002/9780470191576]

[50]   Cushing, B.L.; Kolesnichenko, V.L.; OConnor, C.J. Recent advances in the liquid-phase syntheses of inorganic nanoparticles. *Chem. Rev.,* **2004**, *104*(9), 3893-3946.
[http://dx.doi.org/10.1021/cr030027b] [PMID: 15352782]

[51]   Livage, J.A.; Henry, M.; Sanchez, C. Sol-gel chemistry of transition metal oxides. *Prog. Solid State Chem.,* **1988**, *18*, 259-341.
[http://dx.doi.org/10.1016/0079-6786(88)90005-2]

[52]   Celerier, S.T.; Laberty, C.; Ansart, F.; Lenormand, P.; Stevens, P. New chemical route based on sol–gel process for the synthesis of oxyapatite $La_{9.33}Si_6O_{26}$. *Ceram. Int.,* **2006**, *32*, 271-276.
[http://dx.doi.org/10.1016/j.ceramint.2005.03.001]

[53]   Celerier, S.; Laberty-Robert, C.; Ansart, F.; Calmet, C.; Stevens, P. Synthesis by sol–gel route of oxyapatite powders for dense ceramics: Applications as electrolytes for solid oxide fuel cells. *J. Eur. Ceram. Soc.,* **2005**, *25*, 2665-2668.
[http://dx.doi.org/10.1016/j.jeurceramsoc.2005.03.197]

[54]   Lu, L.; Zeng, Y. Influence of pH on the property of apatite-type lanthanum silicates prepared by sol-gel process. *J Wuhan Univ Technol Mater Sci Ed,* **2012**, *27*, 841-846.
[http://dx.doi.org/10.1007/s11595-012-0559-3]

[55] Vojisavljević, K.; Chevreux, P.; Jouin, J.; Malič, B. Characterization of the alkoxide-based sol-gel derived $La_{9.33}Si_6O_{26}$ powder and ceramic. *Acta Chim. Slov.,* **2014**, *61*(3), 530-541.
[PMID: 25286208]

[56] Li, Q.; Dong, L.M. Fast Solidification and Electrical Conductivity of Apatite-type Nanoceramics. *Adv. Mat. Res.,* **2014**, *981*, 909-913.

[57] Qingle, S.H.; Zhang, H.; Tianjing, L.I.; Fangli, Y.U.; Haijun, H.O.; Pengde, H.A. Preparation and characterization of LSO-SDC composite electrolytes. *J. Rare Earths,* **2015**, *33*, 304-309.
[http://dx.doi.org/10.1016/S1002-0721(14)60418-X]

[58] El Nahrawy, A.M.; Afify, H.H.; Ali, A.I. Investigations of structural and spectroscopic characterization of lanthanum silicate nanocrystalline. *Int J Adv Eng Technol Comput Sci,* **2014**, *1*, 28-35.

[59] Shen, J; Ding, X; Gao, X; Wu, H; Wang, G J *Effect of La deficiency on electrical properties of lanthanum silicate for intermediate temperature solid oxide fuel cells.,*

[60] Yang, T.; Zhao, H.; Han, J.; Xu, N.; Shen, Y.; Du, Z.; Wang, J. Synthesis and densification of lanthanum silicate apatite electrolyte for intermediate temperature solid oxide fuel cell *via* co-precipitation method. *J. Eur. Ceram. Soc.,* **2014**, *34*, 1563-1569.
[http://dx.doi.org/10.1016/j.jeurceramsoc.2013.12.007]

[61] Jo, S.H.; Muralidharan, P.; Kim, D.K. Low-temperature sintering of dense lanthanum silicate electrolytes with apatite-type structure using an organic precipitant synthesized nanopowder. *J. Mater. Res.,* **2009**, *24*, 237-244.
[http://dx.doi.org/10.1557/JMR.2009.0018]

[62] Kharlamova, T.; Vodyankina, O.; Matveev, A.; Stathopoulos, V.; Ishchenko, A.; Khabibulin, D.; Sadykov, V. The structure and texture genesis of apatite-type lanthanum silicates during their synthesis by co-precipitation. *Ceram. Int.,* **2015**, *41*, 13393-13408.
[http://dx.doi.org/10.1016/j.ceramint.2015.07.128]

[63] Pechini, M.P. US Pat., 3 330 697, **1967**.

[64] Segal, D. Chemical synthesis of ceramic materials. *J. Mater. Chem.,* **1997**, *7*, 1297-1305.
[http://dx.doi.org/10.1039/a700881c]

[65] Kioupis, D.; Argyridou, M.; Gaki, A.; Kakali, G. Wet chemical synthesis of $La_{9.83-x}Sr_xSi_6O_{26+\delta}$ ($0 \leq x \leq 0.50$) powders, characterization of intermediate and final products. *J. Rare Earths,* **2015**, *33*, 320-326.
[http://dx.doi.org/10.1016/S1002-0721(14)60420-8]

[66] Jothinathan, E.; Vanmeensel, K.; Vleugels, J.; Van der Biest, O. Powder synthesis, processing and characterization of lanthanum silicates for SOFC application. *J. Alloys Compd.,* **2010**, *495*, 552-555.
[http://dx.doi.org/10.1016/j.jallcom.2009.10.106]

[67] Sōmiya, S.; Roy, R. Hydrothermal synthesis of fine oxide powders. *Bull. Mater. Sci.,* **2000**, *23*, 453-460.
[http://dx.doi.org/10.1007/BF02903883]

[68] Lencka, M.M.; Riman, R.E. Thermodynamics of the hydrothermal synthesis of calcium titanate with reference to other alkaline-earth titanates. *Chem. Mater.,* **1995**, *7*, 18-25.
[http://dx.doi.org/10.1021/cm00049a006]

[69] Nishizawa, H.; Yamasaki, N.; Matsuoka, K.; Mitsushio, H. Crystallization and transformation of zirconia under hydrothermal conditions. *J. Am. Ceram. Soc.,* **1982**, *65*, 343-346.
[http://dx.doi.org/10.1111/j.1151-2916.1982.tb10467.x]

[70] Noviyanti, A.R.; Prijamboedi, B.; Marsih, I.N.; Ismu, I. Hydrothermal Preparation of Apatite-Type Phases $La_{9.33}Si_6O_{26}$ and $La_9M_1Si_6O_{26.5}$ (M= Ca, Sr, Ba). *J Math Fund Sci,* **2012**, *44*, 193-203.

[71]   Kitamura, N; Kaneko, K; Idemoto, Y. Low-Temperature Synthesis and Study of Apatite-Type Lanthanum Silicates. *InMeeting Abstracts,* **2012**, *2*, 48-48. The Electrochemical Society.

[72]   Su, B. Novel fabrication processing for improved lead zirconate titanate (PZT) ferroelectric ceramic materials (Doctoral dissertation, University of Birmingham, IRC in Materials for High Performance Applications, School of Metallurgy and Materials, Faculty of Engineering.).

[73]   Karthikeyan, J.; Berndt, C.C.; Tikkanen, J.; Reddy, S.; Herman, H. Plasma spray synthesis of nanomaterial powders and deposits. *Mater. Sci. Eng. A,* **1997**, *238*, 275-286. [http://dx.doi.org/10.1016/S0921-5093(96)10568-2]

[74]   Gao, W; Zhang, C; Lapostolle, F; Liao, H; Coddet, C; Ji, V. Synthesis of Lanthanum Silicates with Apatite-type Structure by Atmospheric Plasma Spraying. *Thermal Spray 2007: Global Coating Solutions (ASM International),* **2007**, 756-759.

[75]   Porras-Vázquez, J.M.; Losilla, E.R.; León-Reina, L.; Marrero-López, D.; Aranda, M.A. Microstructure and Oxide Ion Conductivity in a Dense $La_{9.33}(SiO_4)_6O_2$ Oxy-Apatite. *J. Am. Ceram. Soc.,* **2009**, *92*, 1062-1068. [http://dx.doi.org/10.1111/j.1551-2916.2009.03032.x]

[76]   Cao, W. *Synthesis of Nanomaterials by High Energy Ball Milling*; Hawk's Perch Technical Writing LLC: Washington, USA, **2014**.

[77]   Rodriguez-Reyna, E.; Fuentes, A.F.; Maczka, M.; Hanuza, J.; Boulahya, K.; Amador, U. Structural, microstructural and vibrational characterization of apatite-type lanthanum silicates prepared by mechanical milling. *J. Solid State Chem.,* **2006**, *179*, 522-531. [http://dx.doi.org/10.1016/j.jssc.2005.11.008]

[78]   Macedo, G.L.; Macedo, D.A.; Rajesh, S.; Martinelli, A.E.; Figueiredo, F.M.; Marques, F.; Nascimento, R.M. Electrical Properties of Lanthanum Silicate Apatite Electrolytes Prepared by an Innovative Chemical Route. *ECS Trans.,* **2014**, *61*, 23-31. [http://dx.doi.org/10.1149/06136.0023ecst]

[79]   Yamagata, C.; Elias, D.R.; Paiva, M.R.; Misso, A.M.; Castanho, S.R. Facile preparation of apatite-type lanthanum silicate by a new water-based sol–gel process. *Mater. Res. Bull.,* **2013**, *48*, 2227-2231. [http://dx.doi.org/10.1016/j.materresbull.2013.02.041]

[80]   Misso, A.M.; Elias, D.R.; Santos, F.D.; Yamagata, C. Low temperature synthesis of lanthanum silicate apatite type by modified sol-gel process. *Adv. Mat. Res.,* **2014**, *975*, 143-148.

[81]   Elias, D.R. *Síntese e caracterização de pós de silicato de lantânio tipo apatita para eletrólito em SOFC*; Universidade de São Paulo, Instituto de Pesquisas Energéticas e Nucleares: São Paulo, **2013**. dissertação[acess 2016-04-25].

[82]   Tian, C.; Liu, J.; Cai, J.; Zeng, Y. Direct synthesis of $La_{9.33}Si_6O_{26}$ ultrafine powder *via* sol–gel self-combustion method. *J. Alloys Compd.,* **2008**, *458*, 378-382. [http://dx.doi.org/10.1016/j.jallcom.2007.03.128]

[83]   Li, B.; Liu, J.; Hu, Y.; Huang, Z. Preparation and characterization of $La_{9.33}Si_6O_{26}$ powders by molten salt method for solid electrolyte application. *J. Alloys Compd.,* **2011**, *509*, 3172-3176. [http://dx.doi.org/10.1016/j.jallcom.2010.10.215]

[84]   Celerier, S.; Laberty-Robert, C.; Ansart, F.; Calmet, C.; Stevens, P. Synthesis by sol–gel route of oxyapatite powders for dense ceramics: Applications as electrolytes for solid oxide fuel cells. *J. Eur. Ceram. Soc.,* **2005**, *25*, 2665-2668. [http://dx.doi.org/10.1016/j.jeurceramsoc.2005.03.197]

CHAPTER 5

A Review on Synthesis Methods of Functional SOFC Materials

Flávia de Medeiros Aquino[1,*], Patrícia Mendonça Pimentel[2,*] and **Dulce Maria de Araújo Melo[3]**

[1] *Department of Renewable Energy Engineering, Federal University of Paraíba, João Pessoa, PB, 58051-900, Brazil*

[2] *Federal Rural University of the Semi-Arid, Angicos, RN, 59515-000, Brazil*

[3] *Department of Chemistry, Federal University of Rio Grande do Norte, Natal, 49100-000, Brazil*

Abstract: Many properties of ceramic materials depend largely on its composition and structure, being also affected by the synthesis method. Intense research activity has been carried out for the development of materials and processes to components of solid oxide fuel cells (SOFCs). Regarding ceramic powders preparation to be used as SOFC components, chemical approaches such as polymeric precursor method, sol-gel, combustion, and co-precipitation have been mentioned in literature. This chapter aims to present some methods that have been used to prepare functional SOFC materials, highlighting its main advantages and disadvantages in the performance of these materials.

Keywords: Ceramic materials, Co-precipitation, Combustion, Fuel cells, Gelatin, Polymeric complexing method, SOFCs, Sol-gel, Solid state reaction, Synthesis methods.

INTRODUCTION

Several technologies are used for the development of SOFCs and currently much we advance in the process of preparing the components that make up the cell: cathode, anode and electrolyte. Technologically, the use of these cells lies some limitations related to materials selection and processing. This fact is mainly due to the high temperatures usually employed, which accelerates corrosion processes, thermal stresses and fatigue of components. In addition, it is essential to exist a

* **Corresponding authors Flávia de Medeiros Aquino and Patrícia Mendonça Pimentel:** Department of Renewable Energy Engineering, Federal University of Paraíba, Brazil; Tel/Fax: +55 83 3216-7268; E-mail: flavia@cear.ufpb.br; Federal Rural University of the Semi-Arid, Angicos, RN, 59515-000, Brazil; Tel/Fax: +55 84 3317 - 8255; E-mail: pimentelmp@ufersa.edu.br

Moisés R. Cesário & Daniel A. de Macedo (Eds.)

good compatibility between electrode (anode and cathode), electrolyte and interconnector materials. Accordingly, the scientific community has been studying and trying to develop materials and processes that may allow the production and effective industrial use of the SOFCs.

The choice of composition, doping concentration and processing conditions during the preparation of a ceramic material must be made taking into consideration several characteristics of the final product since all these parameters influence the electrical conductivity, as indicated in Fig. (**1**). In an ideal ceramic processing is expected that the obtained material presents: high chemical purity, chemical homogeneity, high sinterability, good control of particle size distribution and reduced particle size. For the processing of advanced ceramic materials various physical and chemical methods have been reported in the literature, such as:

- Physical or Ceramic methods
    1. Solid state reaction (mixture of oxides)
    2. Combustion synthesis

- Chemical Methods
    1. Sol-gel
    2. Polymeric complexing method
    3. Modified Polymeric complexing method – Use of gelatin
    4. Co-precipitation

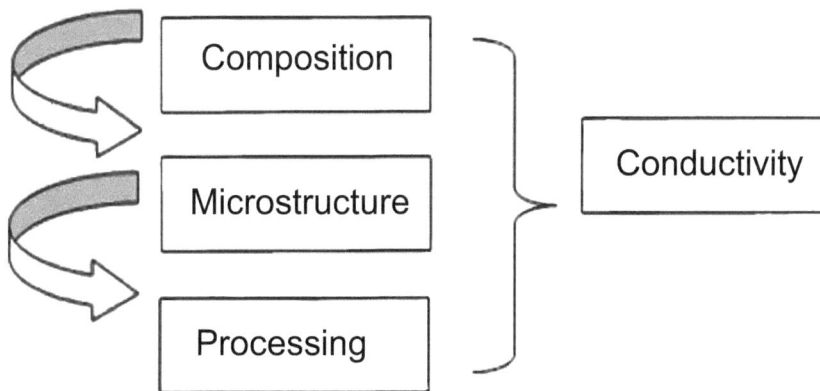

Fig. (1). Correlation between microstructure, composition, processing and conductivity of ceramic materials.

The development of new functional materials and advanced manufacturing techniques are fundamental in reducing the SOFC operating temperature, and as a result, decrease the costs to fabricate these electrochemical devices. Furthermore, the optimization of synthesis techniques are important for the cost reduction during the manufacturing. With this in mind, the flowchart in Fig. (2) shows the procedure usually adopted to synthesize particulate materials.

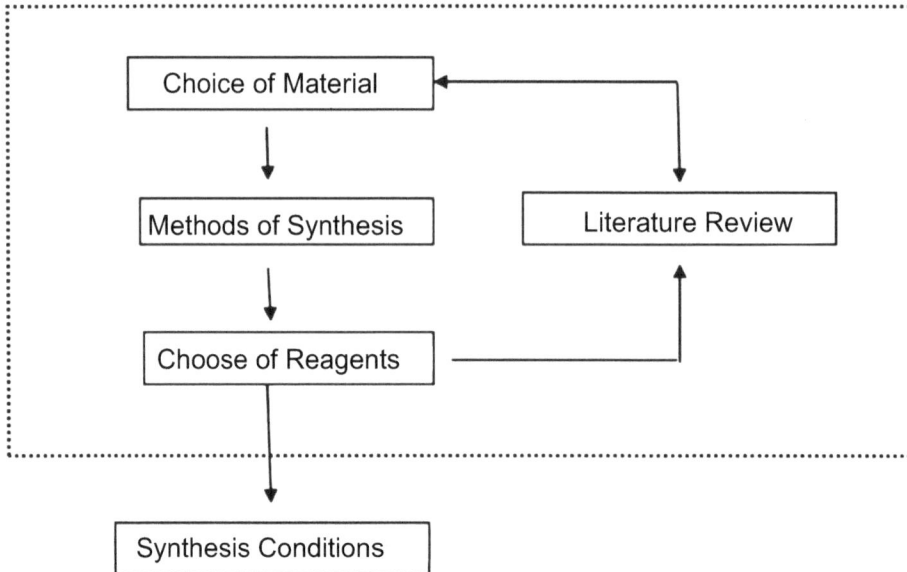

Fig. (2). Steps of the procedure usually adopted to prepared ceramic powders.

TYPES OF SYNTHESIS METHODS

Solid State Reaction (Mixture of Oxides)

The solid state reaction also known as conventional ceramic route is a synthesis method based on solid state diffusion, being characterized by the mechanical mixture of oxides and/or metal carbonates of interest *via* milling followed by heat treatment. In addition to oxides and carbonates other such as chlorides, nitrates, acetates may be used. It's a simple process, relatively inexpensive and is the most widely used industrially. In general, this method requires high temperatures and long calcining times to obtain the desired phases. The reproducibility of the method is limited, mainly due to large particle size distribution and material loss resulting of high temperature volatility. Below are summarized some disadvantages and advantages of this method.

Advantages:

1. Easy implementation and execution;
2. Production of large powder quantity (industrial method);

Disadvantages:

1. High contamination with grinding balls;
2. Requires high calcining temperatures;
3. Production of micrometer scale materials;
4. Poor stoichiometric control, limited by the diffusion kinetics.

The physical principles taking place during a solid state reaction are solid state diffusion, nucleation and grain growth. Many important industrial processes and reactions in the treatment of materials depend on the mass transport of solid, liquid or gaseous species (at the microscopic level) in another solid phase. This is necessarily accomplished by diffusion which can be defined as being the mechanism by which the mass is transported through materials. Whereas atoms in gases, liquids and solids are in constant motion and migrate over time, diffusion also can be defined as the migration of atoms from one place to another crystal lattice or transport of matter in the solid state by atomic motion induced by thermal agitation.

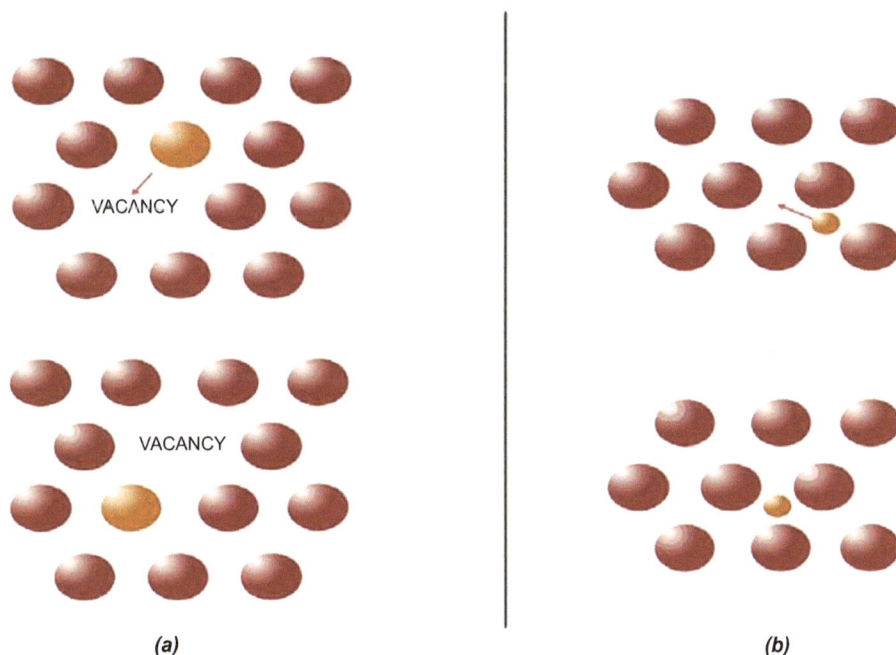

Fig. (3). Schematic representations of (**a**) diffusion gap and (**b**) interstitial diffusion.

In solid state, atoms are constantly moving and changing their positions. Two conditions are essential for atoms to do these moviments: (1) It is necessary that there is a vacancy on the adjacent site and (2) atoms should have enough energy to break the atomic bonds in crystal lattice and to provokes some distortion of the crystal lattice during its displacement, as shown in Fig. (**3**). This energy has vibrational nature. At a certain temperature, a short fraction of the total number of atoms is able to moving by diffusion, owing to their magnitudes of the vibrational energy. This fraction rises with increasing temperature.

Experimental Procedure of Solid State Reaction

The powder preparation by solid state reaction constitutes the traditional mechanical mixing of starting materials, which can be done in two ways: homogenizing in a mortar or by grinding in mills, in this case there is improved homogeneity and reduction of grain size. The difference between them is basically the total manufacturing time.

A new cathode intermediate temperature solid oxide fuel cells, $BaBi_{0.05}Co_{0.8}Nb_{0.15}O_{3-d}$ (BBCN) using the solid reaction method have reported by Zhou *et al.* [1]. Precursor powder was calcined at 850 and 1100 °C for 10 h aim to obtain the perovskite phase. BBCN powders calcined at 1100 °C were single phase with cubic perovskite-type structure. The authors performed an electrochemical stability test of the BBCN cathode under real fuel cell conditions using LSGM electrolyte. BBCN was chemically compatible with the electrolyte at 1000 °C for 5 h. The following flowchart (Fig. **4**) illustrates the process of synthesis of $BaBi_{0.05}Co_{0.8}Nb_{0.15}O_{3-d}$ by solid state reaction [1].

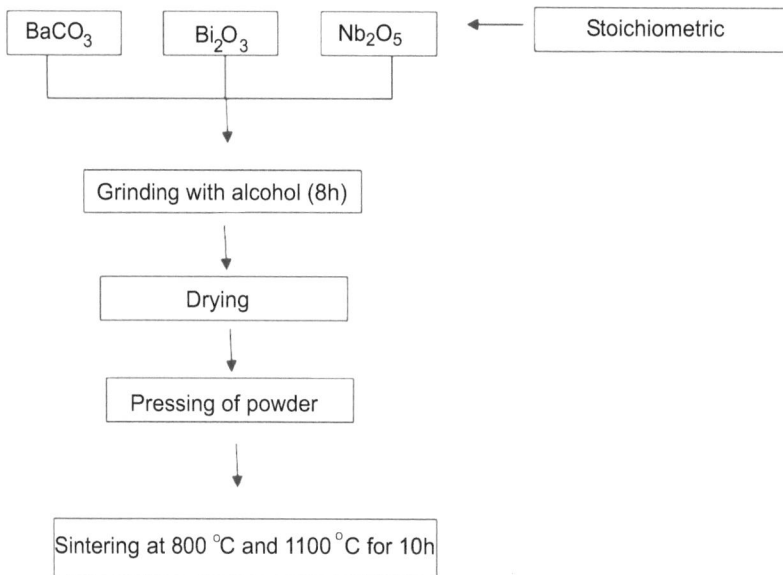

Fig. (4). Flowchart of a typical experimental procedure used for solid state reactions.

Raghvendraa *et al.* [2] studied the effect of calcia doping on proprieties of $La_{0.9}Sr_{0.1}Ga_{0.8}Mg_{0.2}O_{2.85}$ (LSGM) solid solutions. The powders were synthesized by conventional solid state route and sintered at 1400 °C for 12 h to obtain the LSGM phase. Few secondary phases were identified along with LSGM phase. The porosity and surface area were suitable for use as SOFC electrolytes.

Combustion Synthesis

Combustion synthesis is a process that uses a combustion reaction to produce inorganic, solid materials, in a redox reaction. Combustion synthesis can be a pure redox reaction between two reducts, but it is also possible to add fuels or oxidizers, *e.g.* nitrate as oxidizer and sucrose, citric acid, urea or glycine as fuel [3]. In the solution combustion process, the most used fuels are glycine and urea. The glycine and urea advantage is they can form stable complexes with metal ions to augment solubility and avoid selective precipitation of the metal ions during water removal. The ash resulting from combustion are composed of particles very fine [4]. The process of combustion synthesis is generally characterized by highly exothermic reactions, with the reaction temperature being on the order of 500 to 4000 K [5].

In the combustion process, usually is used as external heat source muffle, plate or microwave oven (microwave assisted self-combustion synthesis). In microwave oven, the heat dissipation is performed directly inside the material and allows a much more uniform temperature distribution than in a conventional oven. Whatever the heating source used, CO_2, H_2O and N_2 are stable gaseous products. The theory of combustion synthesis is based on the propellants chemistry. Jain *et al.* [6] introduced a method to calculate the oxidizing and reducing reaction character, with carbon and hydrogen being considered reducing element with valences +4 and +1, respectively; oxygen as an oxidizer element with valence -2 and nitrogen is considered a zero valence.

Advantages:

1. Synthesis of oxide powders with compositional homogeneity in very short times;
2. No high temperature calcination is needed.

Disadvantage:

1. Easy powder agglomeration together with undesired phases.

Experimental Procedure of the Combustion Synthesis

The synthesis of $Sm_{0.5}Sr_{0.5}CoO_{3-d}$ perovskite as a cathode material for IT-SOFCs is given as an example to demonstrate the combustion synthesis [7]. The starting materials used in the synthesis were $Sm(NO_3)_3 \cdot 6H_2O$, $La(NO_3)_3 \cdot 6H_2O$, $Sr(NO_3)_2$, $Co(NO_3)_2 \cdot 6H_2O$, and glycine (NH_2CH_2COO). The solution–combustion process is shown in Fig. (**5**). Metal nitrates are employed both as metal precursors and oxidizing agents. The metal nitrates were dissolved in deionized water. A calculated amount of the amino acid glycine (0.7 mol of NO_3^-) was also dissolved in deionized water. The glycine solution was slowly added to the metal nitrate aqueous solution under constant stirring. Glycine acts as a complexing agent for metal cations of varying sizes as it has a carboxylic group at one end and an amino group at the other end. The resulting solution was heated until achieve a concentration of about 2 mol.l^{-1} on metal nitrate basis. While the solution was still hot, it was added drop wise to a 2 L glass beaker that was preheated between 300 and 400 °C. The water in the solution quickly evaporated the resulting viscous liquid swelled, auto-ignited and started a highly exothermic self-contained combustion process, converting the precursor materials into a fine powder of oxides. Oxygen from air does not play a relevant role during the combustion process. The overall combustion reactions can be represented as [7]:

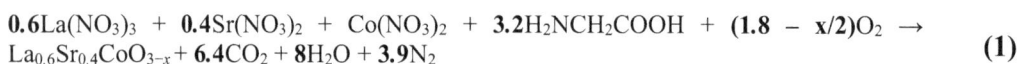

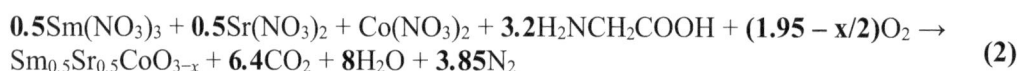

$$\mathbf{0.6}La(NO_3)_3 + \mathbf{0.4}Sr(NO_3)_2 + Co(NO_3)_2 + \mathbf{3.2}H_2NCH_2COOH + (\mathbf{1.8} - x/2)O_2 \rightarrow$$
$$La_{0.6}Sr_{0.4}CoO_{3-x} + \mathbf{6.4}CO_2 + \mathbf{8}H_2O + \mathbf{3.9}N_2 \tag{1}$$

$$\mathbf{0.5}Sm(NO_3)_3 + \mathbf{0.5}Sr(NO_3)_2 + Co(NO_3)_2 + \mathbf{3.2}H_2NCH_2COOH + (\mathbf{1.95} - x/2)O_2 \rightarrow$$
$$Sm_{0.5}Sr_{0.5}CoO_{3-x} + \mathbf{6.4}CO_2 + \mathbf{8}H_2O + \mathbf{3.85}N_2 \tag{2}$$

The resulting black powder containing carbon residue was further calcined to obtain the desired product. Small portions of this powder were heat treated in air at various temperatures between 700 and 1300 °C for 2 h to study the crystalline phases evolution and the oxides obtained.

Several modifications of combustion method have been reported. Conceição *et al.* [8] have investigated the different fuels effects (urea, glycine, citric acid, and sucrose) and sintering temperature on the porosity and electrical conductivity of $La_{0.7}Sr_{0.3}Co_{0.5}Fe_{0.5}O_3$. A furnace previously heated at 400 °C was used as heat source. The powders were calcined at 750 °C for 10 h and sintered at temperatures of 950 –1100 °C for 4 h. The authors observed the formation of minor secondary phases in the powders synthesized using urea and sucrose as fuels. The synthesis using citric acid did not present any ignition. The combustion reaction was more complete with glycine.

```
┌─────────────────────────┐        ┌─────────────────────────┐
│   Nitrates of La, Sm, Sr,│        │     Glycine + water     │
│        Co + water       │        │                         │
└─────────────────────────┘        └─────────────────────────┘
                    │                          │
                    └──────────────┬───────────┘
                                   ▼
                  ┌───────────────────────────────┐
                  │   Mix metal nitrates and      │
                  │  glycine solutions under      │
                  │           stirring            │
                  └───────────────────────────────┘
                                   │
                                   ▼
                  ┌───────────────────────────────┐
                  │ Clear red solution; heat at ~ 80 °C;│
                  │  concentrate to ~ 2 M metal nitrate │
                  │              basis            │
                  └───────────────────────────────┘
                                   │
                                   ▼
                  ┌───────────────────────────────┐
                  │     Add above solution        │
                  │   dropwise to a beaker        │
                  │  preheated to 300-400°C       │
                  └───────────────────────────────┘
                                   │
                                   ▼
            ┌──────────────────────────────────────────┐
            │  Black powder, heat treat 700 - 1300°C,   │
            │            2h each, in air                │
            └──────────────────────────────────────────┘
```

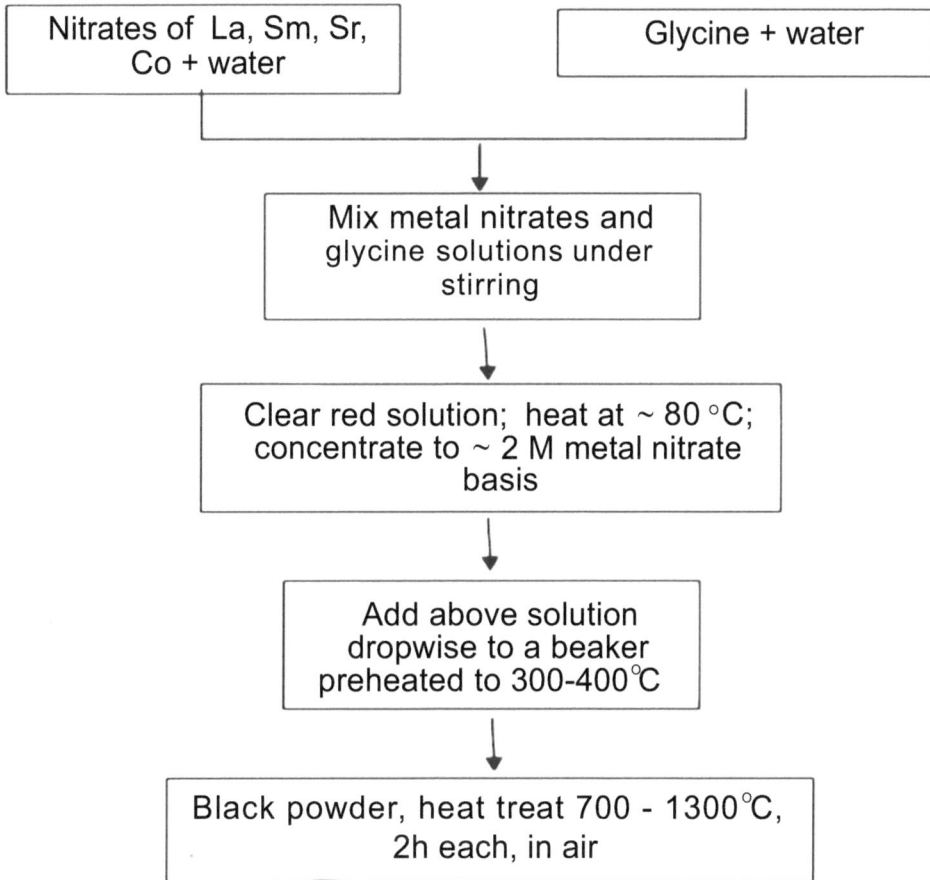

Fig. (5). Flowchart for solution–combustion synthesis of $Sm_{0.5}Sr_{0.5}CoO_{3-x}$ and $La_{0.6}Sr_{0.4}CoO_{3-x}$ nanopowders [7].

Silva *et al.* [9] have prepared lanthanum strontium manganite (LSM) and chromite (LSC) by combustion method using urea and glycine as fuels. The authors used a muffle furnace previously heated at 300 °C. The precursor powders were calcined at 600 °C for 6 h. It was observed secondary phase formation in the powders synthesized with glycine.

Molero-Sáncheza *et al.* [10] shown that the microwave-based combustion synthesis can lower the material manufacturing expense and enhance the cathode performance towards SOFC applications. The authors synthesized $La_{0.3}Ca_{0.7}Fe_{0.7}Cr_{0.3}O_{3-\delta}$ (LCFCr) powders by three different methods: combustion, microwave-assisted combustion, and microwave-assisted sol–gel synthesis. Microwave methods were beneficial to rise the surface area and to decrease the total synthesis time. Based on their results, the microwave-assisted combustion method was elected the preferred chemical route.

Sol-gel

The sol-gel method is characterized by the transition from a sol system to a gel system. The term sol is used to define a colloidal particles dispersion (size between 1 and 100 nm) stable in a fluid, while the term gel refers to a system formed by the rigid structure of colloidal particles (colloidal gel) or polymer chains (polymer gel) that immobilize the liquid phase. Drying of the colloidal suspension under normal conditions of pressure and temperature leads to the formation of the gel.

Advantages:

1. Easy implementation and execution;
2. Low temperature synthesis;
3. Nano-scale materials;
4. Good stoichiometric control.

Disadvantages:

1. Powder agglomeration;
2. Low solution stability.

The sol-gel process can be divided into two classes depending on the nature of the inorganic precursors used: salts (chlorides, nitrates, sulfates, *etc.*) or alkoxides. Those routes involving the use of alkoxides are the most versatile.

Experimental Procedure of Sol-gel Method

$PrBa_{0.5}Sr_{0.5}Co_{2-x}Ni_xO_{5-\delta}$ (x = 0.1, 0.2 and 0.3) cathode materials were prepared by combining EDTA and citrate sol-gel methods. Cathodes supported on YSZ electrolytes had polarization resistances of about 0.297 $\Omega.cm^2$ and maximum power density of 120 mW at 800 °C [11]. The following flowchart (Fig. **6**) illustrates the process sol-gel.

Njoku *et al.* [12] synthesized a novel $Ce_{0.8}Sm_{0.2}Fe_{0.9}Ir_{0.03}Co_{0.07}O_{3-\delta}$ cathode material using the sol gel method. Powders were calcined at 800 – 1000 °C for 10 h. Secondary phases of Fe_2O_3, $FeSmO_2$ and IrO_2 were detected in all samples. SEM images showed porous agglomerates. The results of electrochemical tests revealed that the powder calcined at 1000 °C exhibited the most promising performance.

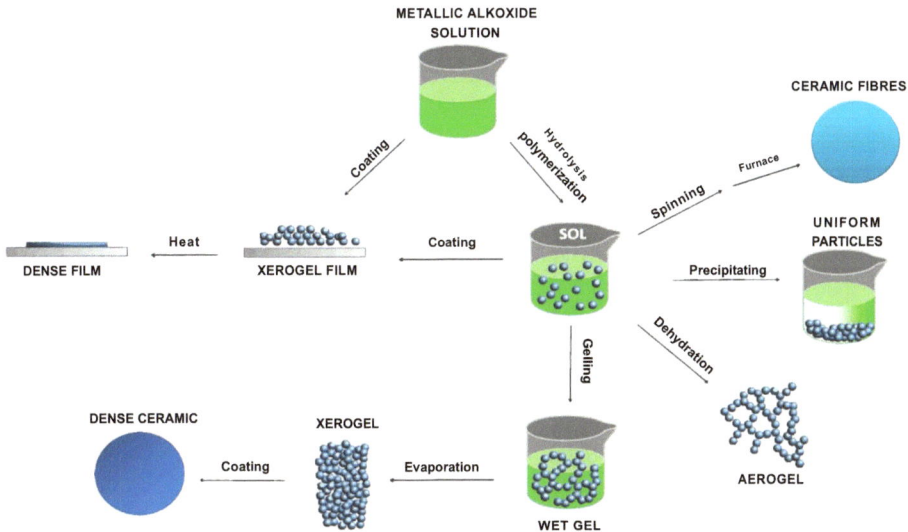

Fig. (6). Flowchart of a typical sol-gel synthesis.

Polymeric Complexing Method

The polymeric precursor method, developed by Pechini [13], is a variation of the sol-gel method. This method consists on formation of a polymeric resin produced by a polyesterification reaction between a metal complex and a polyhydroxy alcohol such as ethylene glycol. Sources of metal cations can be nitrates, oxides, carbonates, acetates, *etc*. The polymer resin can be used to prepare thin films or powders. To obtain the powder, the resin is calcined at a temperature of about 400 °C to break the polymer and removal of organic matter through the release of gases (H_2O, CO, CO_2). The result is a dark and fragile material. Subsequent calcination steps at higher temperatures leads to the formation of fine oxides with the desired stoichiometry. The chemical reaction between citric acid and chelating metal ions is shown in Fig. (**7**).

Advantages:

1. Easy deployment;
2. Low contamination;
3. Low temperature synthesis;
4. No diffusion problem;

Disadvantages:

1. Expensive reagents;

2. Powders clusters;
3. Low solution stability.

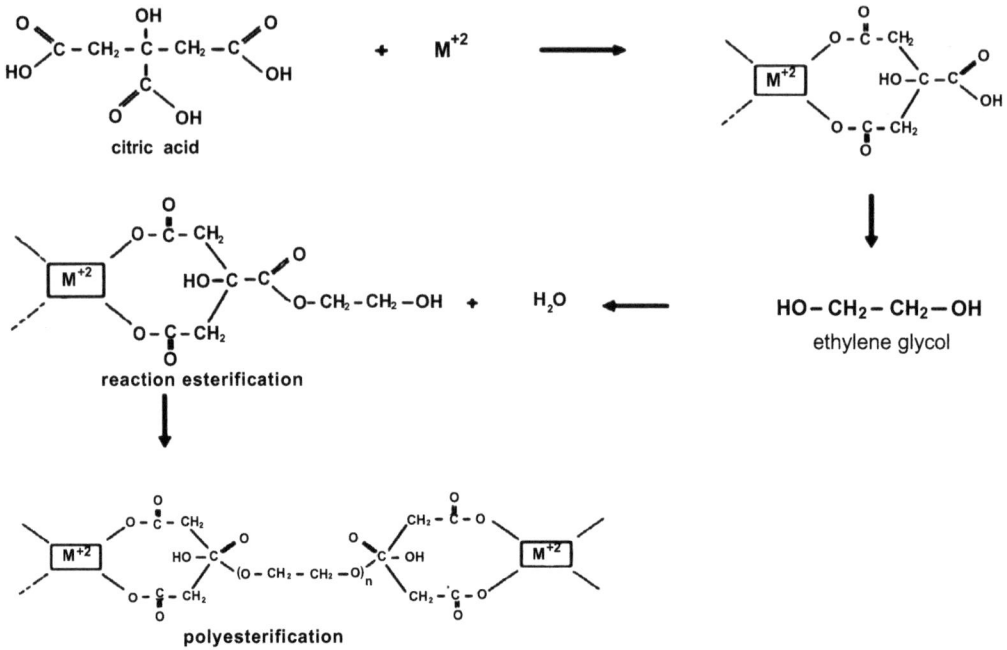

Fig. (7). Mechanism of chemical reactions for the polymeric complexing method.

Experimental Procedure of Polymeric Complexing Method

$La_{1-x}Sr_xMnO_3$ powders were synthesized by the polymeric precursor route (Pechini method) using lanthanum nitrate, strontium nitrate, manganese nitrate, citric acid, ethylene glycol, and distilled water as starting materials [7]. The 'nitrate solutions concentrations were previously determined by gravimetry. Manganese citrate solution was prepared from $Mn(NO_3)_2.4H_2O$ and citric acid, using the citric acid/metal cation molar ratio of 3:1, under stirring at 70 °C for 2 h. The network modifier and the dopant were later added. The resulting solution was subjected to stirring and heating at 90 °C. Afterwards, ethylene glycol was added at a 60:40 citric acid/ethylene glycol mass ratio. This solution was kept at mechanical agitation and heating at 95 °C for 2 h to form the polymeric resin, which was later heat-treated at 300 °C for 2 h, yielding a fragile and porous solid. Such material was initially disaggregated in a mortar and later ground for 2 h in a planetary mill. The powders were calcined at 450, 700 and 950 °C for 4 h using heating rate of 5 °C/min [7]. Fig. (**8**) illustrates the process of synthesis.

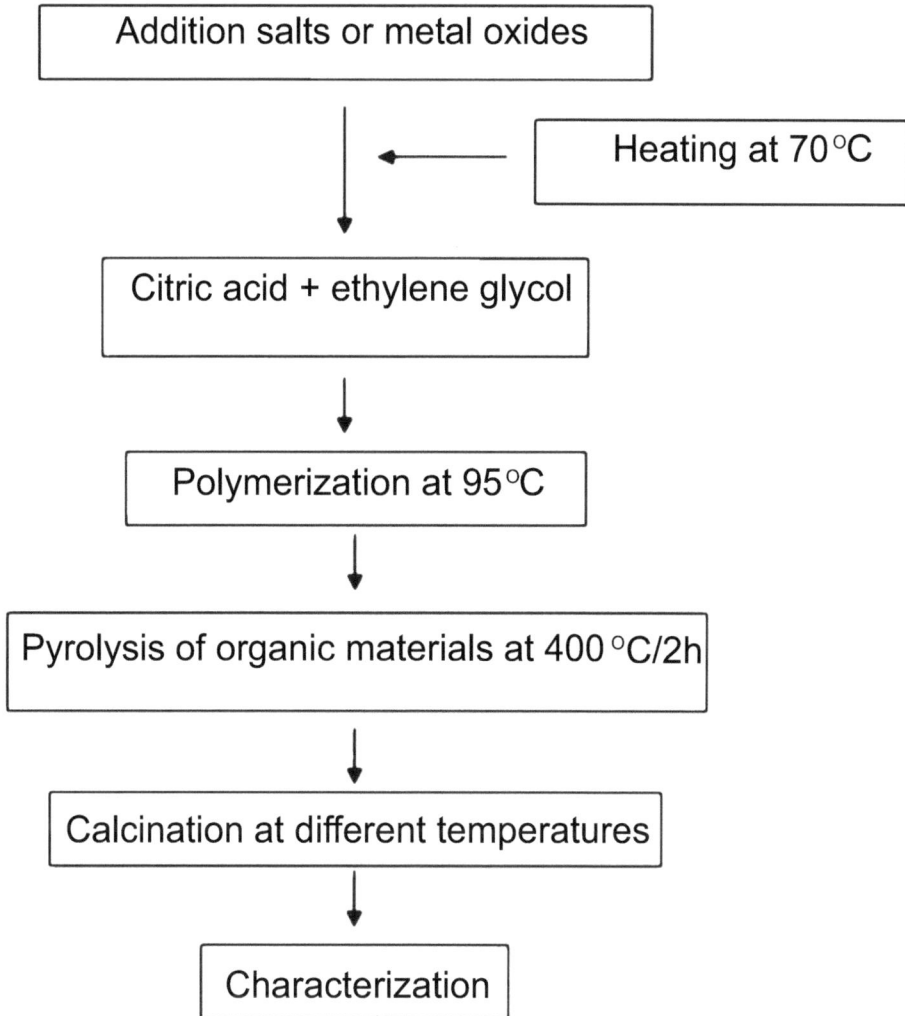

Fig. (8). Flowchart of a typical polymeric complexing synthesis.

Niwa *et al.* [14] have prepared $LaNi_{1-x}FexO_3$ ($1.0 \geqq x \geqq 0.4$) powders by Pechini method. The powders were compared with materials of same composition synthesized by state solid reaction. The powders calcined at 750 °C were single phase, whereas large quantity of raw materials remained in solid state reacted material. The powders of composition $LaNi_{0.6}Fe_{0.4}O_3$ showed superior properties as cathode material.

Modified Polymeric Complexing Method

Several modifications of the sol-gel and Pechini methods have been reported in literature. Some of these have several disadvantages, such as high cost of raw materials, time consuming procedures, and use of hazardous organic compounds. A method in which the alkoxides are replaced by gelatin was developed by Monrós *et al.* [15] and has undergone changes by some researchers aiming to reduce the number of steps and the total process time [12, 16, 17].

Gelatin is the denatured collagen protein resulted by boiling collagen with water. Collagen is the main skin protein, being primarily constituted of glycine, proline and hydroxyproline. Gelatin has several protein chains composed of aminoacids. These form large chains that can chelate cations through their groups amino and carbonyl [15].

The procedure consists on forming colloidal dispersions between gelatin and water, adding metallic ions and providing heat in this dispersion to reduce the volume up to the formation of a gel, which is subsequently calcined at a predetermined temperature for decomposition of gelatin and formation of inorganic oxides [12, 18]. Fig. (**9**) summarizes the process.

The synthesis using gelatin is simpler than sol-gel or Pechini methods for several reasons: the raw materials are less expensive than citric acid and ethylene glycol; the steps and time of processing are reduced. The process shows a higher formal functionality of gelatin molecule for the chelate effect when compared to the citric acid from Pechini method. Gelatin forms a lyophilic gel structure, avoiding common condensation step of Pechini method [15].

Advantage:

1. Gelatin is an inexpensive and non-toxic material;
2. Low synthesis temperature;

Disadvantage:

1. Formation of agglomerated particles.

Experimental Procedure of the Modified Polymeric Complexing Method

A modified Pechini method was used to synthesize cathode powders the fuel cell using gelatin and ($Ba(NO_3)_2$, $Sr(NO_3)_2$, $Co(NO_3)_2.6H_2O$, $Fe(NO_3)_3.9H_2O$, $Nd(NO_3)_3.6H_2O$ and $Sm(NO_3)_3.6H_2O$), as starting materials. Gelatin was added to water at 70 °C and stirred for about 40 min. Afterwards, nitrates were added to the beaker under constant stirring 70 °C for each 15 min. The precursor was kept at

90 °C with stirring to evaporate water. The resulting gels were dried in an oven at 350 °C for 2 h. The precursor materials were calcined at 1000 °C for 4 h in air to obtain the perovskite type phase [16].

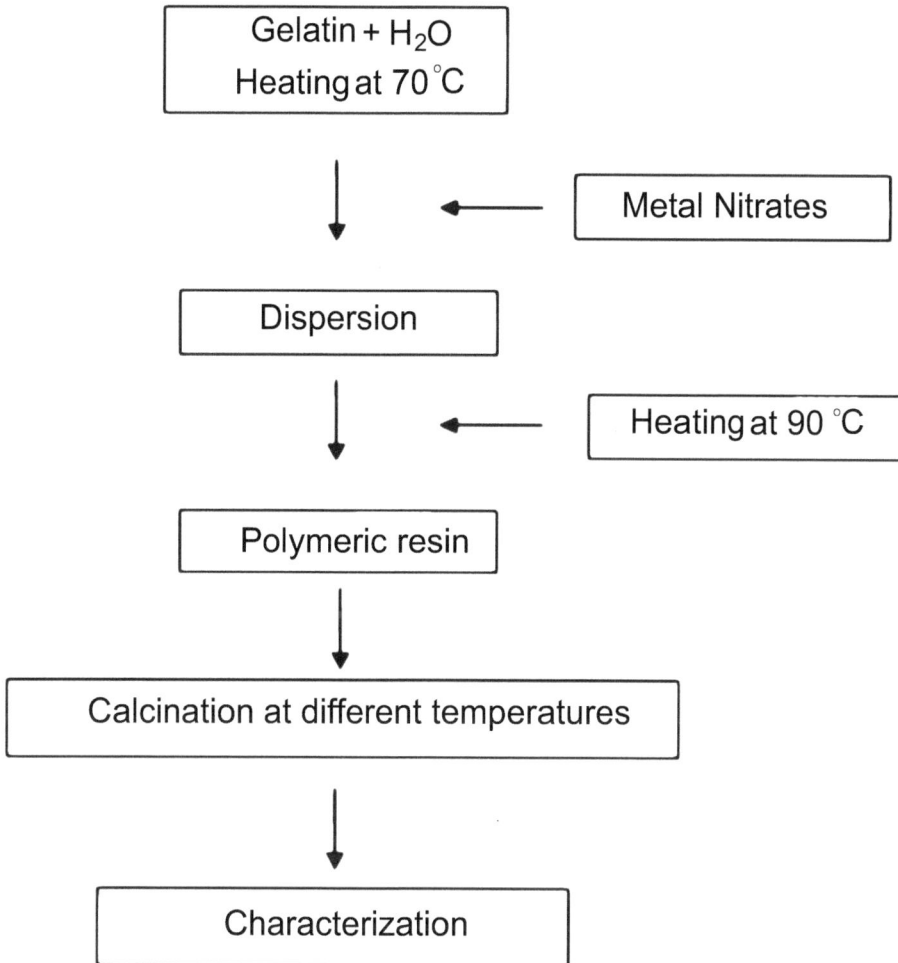

```
         ┌─────────────────────────┐
         │    Gelatin + H₂O        │
         │   Heating at 70 °C      │
         └─────────────────────────┘
                     │                    ┌──────────────────┐
                     │  ◄───────────────  │  Metal Nitrates  │
                     ▼                    └──────────────────┘
         ┌─────────────────────────┐
         │      Dispersion         │
         └─────────────────────────┘
                     │                    ┌──────────────────┐
                     │  ◄───────────────  │ Heating at 90 °C │
                     ▼                    └──────────────────┘
         ┌─────────────────────────┐
         │    Polymeric resin      │
         └─────────────────────────┘
                     │
                     ▼
    ┌──────────────────────────────────────────┐
    │  Calcination at different temperatures    │
    └──────────────────────────────────────────┘
                     │
                     ▼
         ┌─────────────────────────┐
         │    Characterization     │
         └─────────────────────────┘
```

Fig. (9). Flowchart of an experimental procedure.

Co-precipitation Method

The co-precipitation method consists in preparing homogeneous solutions containing metal salts (chlorides, acetates, nitrates, *etc.*) and soluble bases (NaOH, NH_4OH) added in stoichiometric proportion to occur simultaneous precipitation of cations in solution. The precipitation occurs due to the formation of an insoluble compound, it is important that the concentration of cations and anions present in the solution exceeds the solubility product (Ksp). To obtain desired oxide is necessary separate product from supernatant and subsequent calcination. The morphological characteristics of material will be influenced by time, temperature and reaction pH, as well as the thermal treatment for the formation of oxides phases [19].

Advantage:

1. Reproducibility;
2. Large scale production;

Disadvantage:

1. Production of powders with low crystallinity;
2. High dispersion in particle size;
3. High calcining temperatures to obtain the desired phase.

Experimental Procedure of Co-precipitation Method

An example of a co-precipitation synthesis (Fig. **10**) is summarized as following: Solid solutions of $Ce_{1-x}Gd_xO_{2-\delta}$ (x = 0.1-0.3) were synthesized by carbonate co-precipitation method. Cerium nitrate, gadolinium nitrate and ammonium carbonate (co-precipitant medium) were used as starting materials. A stoichiometric mixture of Gd^{3+} and Ce^{3+} ions were prepared by dissolving metallic nitrates in 150 mL distilled water. This solution was added to an ammonium carbonate solution under constant stirring until the formation of a white precipitate (pH 9). The precipitate was filtered and washed several times with distilled water and then with ethanol. The precipitate was dried at 353 K for 12 h and thereafter was calcined at different temperatures [20].

Fang *et al.* [21] have been prepared Ni–Samaria-doped ceria (SDC) for IT-SOFC using a co-precipitation technique. Metallic nitrates were used as precursors and $(NH_4)_2CO_3$ as mineralizer. The precipitates were calcined at 600, 700 and 800 °C for 2 h. NiO-SDC composites were reduced to Ni-SDC cermets by thermal treatment at 750 °C in H_2. The authors also synthesized the cermets by the mechanical mixing method, for comparison, and concluded that conductivity of the cermets synthesized by co-precipitation was remarkably higher than that of mechanical derived materials.

Pelosato *et al.* [22] have reported the synthesis of $LaMnO_3$ (LM), $La_{0.70}Sr_{0.30}MnO_{3-\delta}$ (LSM), $La_{0.80}Sr_{0.20}FeO_{3-\delta}$ (LSF), and $La_{0.75}Sr_{0.25}Cr_{0.50}Mn_{0.50}O_{3-\delta}$ (LSCM) perovskite-type oxides as electrode materials for IT-SOFC. The powders have been prepared using a carbonate co-precipitation method and were calcined at 300, 600, 800,

900, and 1000 °C to obtain the perovskite phase. Temperatures of 1150 °C (LSCM) and 1350 °C (LM, LSM, LSF) also were investigated for oxides. The perovskite phase formation occurred upon thermal treatment at 600 °C in LSF and LSCM oxides and above 800 °C for the other powders. The powders synthesized at 1000 °C were single phase.

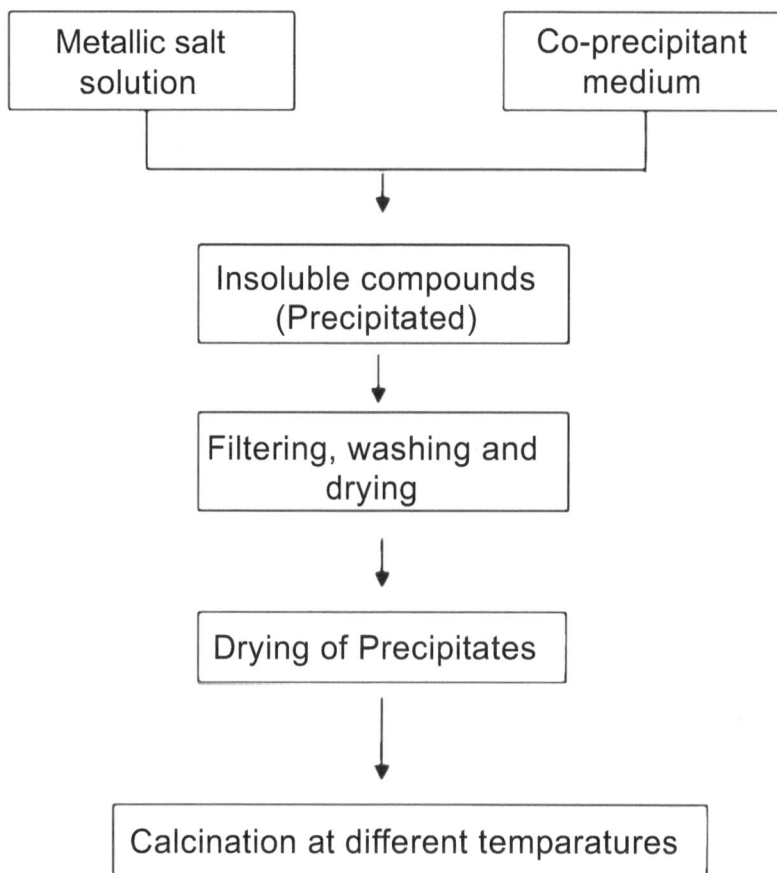

Fig. (10). Flowchart of an experimental procedure.

CONFLICT OF INTEREST

The authors confirm that they have no conflict of interest to declare for this publication.

ACKNOWLEDGEMENTS

The authors thank the cooperation of universities Federal University of Paraíba (UFPB), Federal Rural University of the Semi-Arid (UFERSA) and Federal

University of Rio Grande do Norte (UFRN).

REFERENCES

[1]   Zhou, Q.; Wei, T.; Li, Z.; An, D.; Tong, X.; Ji, Z.; Wang, W.; Lu, H.; Sun, L.; Zhang, Z.; Xu, K. Synthesis and characterization of $BaBi_{0.05}Co_{0.8}Nb_{0.15}O_{3-d}$ as a potential IT-SOFCs cathode material. *J. Alloys Compd.,* **2015**, *627*, 320-323.
        [http://dx.doi.org/10.1016/j.jallcom.2014.11.187]

[2]   Raghvendraa, Singha RK, Sinhab ASK, Singha P. Investigations on structural and electrical properties of calcium substituted LSGM electrolyte materials for IT-SOFC. *Ceram. Int.,* **2014**, *40*, 10711-10718.
        [http://dx.doi.org/10.1016/j.ceramint.2014.03.058]

[3]   Bansal, N.P.; Zhong, Z. Combustion synthesis of $Sm_{0.5}Sr_{0.5}CoO_{3-x}$ and $La_{0.6}Sr_{0.4}CoO_{3-x}$ nanopodews for solid oxide fuel cell cathodes. *J. Power Sources,* **2006**, *158*, 148-153.
        [http://dx.doi.org/10.1016/j.jpowsour.2005.09.057]

[4]   Shao, Z.; Zhou, W.; Zhu, Z. Advanced synthesis of materials for intermediate-temperature solid oxide fuel cells. *Prog. Mater. Sci.,* **2012**, *57*, 804-874.
        [http://dx.doi.org/10.1016/j.pmatsci.2011.08.002]

[5]   Munir, ZA Field-Effects in self-propagating solid-state reactions. *Zeitschrift fur Physikalische chenmie.,* **1998**, *207*, 39-57.
        [http://dx.doi.org/10.1524/zpch.1998.207.Part_1_2.039]

[6]   Jain, S.R.; Adiga, K.C.; Pai Verneker, V.R. A new approach to thermochemical calculations of condensed fuel-oxidizer mixtures. *Combust. Flame,* **1981**, *40*, 71-79.
        [http://dx.doi.org/10.1016/0010-2180(81)90111-5]

[7]   Rabelo, A.A.; Macedo, M.C.; Melo, D.M.; Paskocimas, C.A.; Martinelli, A.E.; Nascimento, R.M. Synthesis and Characterization of $La_{1-x}Sr_xMnO_{3\pm\delta}$ Powders Obtained by the Polymeric Precursor Route. *Mater. Res.,* **2011**, *14*, 91-96.
        [http://dx.doi.org/10.1590/S1516-14392011005000018]

[8]   da Conceição, L.; Silva, A.M.; Ribeiro, N.F.; Souza, M.M. Combustion synthesis of $La_{0.7}Sr_{0.3}Co_{0.5}Fe_{0.5}O_3$ (LSCF) porous materials for application as cathode in IT-SOFC. *Mater. Res. Bull.,* **2011**, *46*, 308-314.
        [http://dx.doi.org/10.1016/j.materresbull.2010.10.009]

[9]   da Silva, A.L.; da Conceição, L.; Rocco, A.M.; Souza, M.M. Synthesis of Sr-doped $LaMnO_3$ and $LaCrO_3$ powders by combustion method: structural characterization and thermodynamic evaluation. *Ceramica,* **2012**, *58*, 521-528.
        [http://dx.doi.org/10.1590/S0366-69132012000400018]

[10]   Molero-Sáncheza, B.; Prado-Gonjalb, J.; Ávila-Brande, D.; Birssa, V.; Morán, E. Microwave-assisted synthesis and characterization of new cathodic material for solid oxide fuel cells: $La_{0.3}Ca_{0.7}Fe_{0.7}Cr_{0.3}O_{3-\delta}$. *Ceram. Int.,* **2015**, *41*, 8411-8416.
        [http://dx.doi.org/10.1016/j.ceramint.2015.03.041]

[11]   Liu, L.; Guo, R.; Wang, S.; Yang, Y.; Yin, D. Synthesis and characterization of $PrBa_{0.5}Sr_{0.5}Co_{2-x}Ni_xO_{5-\delta}$ (x = 0,1, 0,2 and 0,3) cathodes for intermediate temperature SOFCs. *Ceram. Int.,* **2014**, *40*, 16393-16398.
        [http://dx.doi.org/10.1016/j.ceramint.2014.07.144]

[12]   Oliveira, F.S.; Pimentel, P.M.; Oliveira, R.M.; Melo, D.M.; Melo, M.A. Effect of lanthanum replacement by strontium in lanthanum nickelate crystals synthetized using gelatin as organic precursor. *Mater. Lett.,* **2010**, *64*, 2700-2703.
        [http://dx.doi.org/10.1016/j.matlet.2010.08.059]

[13] PECHINI, NUS. Patent, n. 3.330.697 – **1967**.

[14] Niwa, E.; Uematsu, C.; Miyashita, E.; Ohzeki, T.; Hashimoto, T. Conductivity and sintering property of $LaNi_{1-x}Fe_xO_3$ ceramics prepared by Pechini method. *Solid State Ion.,* **2011**, *201*, 87-93. [http://dx.doi.org/10.1016/j.ssi.2011.08.004]

[15] Monrós, G; Carda, J; Tena, MA; Escribano, P; Badenes, J; Cordoncillo, E Spinels from gelatine-protected gels. *J Mater Chem.,* **1995**, *15*, 85-90.

[16] Aquino, F.M.; Marques, F.M.; Melo, D.M.; Macedo, D.; Yaremchenko, A.A.; Figueiredo, F. Preparation of $(Ba,Sr)_{0.5}Sm_{0.5}Co_{0.8}Fe_{0.2}O_{3-\delta}$ and $(Ba,Sr)_{0.5}Nd_{0.5}Co_{0.8}Fe_{0.2}O_{3-\delta}$ cathodes for IT-SOFCs. *Adv Mater Res (Online),* **2014**, *975*, 137-142. [http://dx.doi.org/10.4028/www.scientific.net/AMR.975.137]

[17] de Menezes, A.S.; Remédios, C.M.; Sasaki, J.M.; da Silva, L.R.; Góes, J.C.; Jardim, P.M.; Miranda, M.A. Sintering of nanoparticles of α-Fe_2O_3 using gelatina. *J. Non-Cryst. Solids,* **2007**, *353*, 1091-1094. [http://dx.doi.org/10.1016/j.jnoncrysol.2006.12.022]

[18] Aquino, F.M.; Melo, D.M.; Pimentel, P.M.; Braga, R.M.; Melo, M.A.; Martinelli, A.E.; Costa, A.F. Characterization and thermal behavior of $PrMO_3$ (M = Co or Ni) ceramic materials obtained from gelatin. *Mater. Res. Bull.,* **2012**, *47*, 2605-2609. [http://dx.doi.org/10.1016/j.materresbull.2012.04.078]

[19] Kim, D.K.; Zhang, Y.; Voit, W.; Rao, K.V.; Muhammed, M. Synthesis and characterization of surfactant-coated superparamagnetic monodispersed iron oxide nanoparticles. *J. Magn. Magn. Mater.,* **2001**, *225*, 30-36. [http://dx.doi.org/10.1016/S0304-8853(00)01224-5]

[20] Anjaneya, K.C.; Nayaka, G.P.; Manjanna, J.; Govindaraj, G.; Ganesha, K.N. Preparation and characterization of $Ce_{1-x}Gd_xO_{2-\delta}$ (x = 0.1 – 0.3) as solid electrolyte for intermediate temperature SOFC. *J. Alloys Compd.,* **2013**, *578*, 53-59. [http://dx.doi.org/10.1016/j.jallcom.2013.05.010]

[21] Fang, X.; Zhu, G.; Xia, C.; Liu, X.; Meng, G. Synthesis and properties of Ni–SDC cermets for IT–SOFC anode by co-precipitation. *Solid State Ion.,* **2004**, *168*, 31-36. [http://dx.doi.org/10.1016/j.ssi.2004.02.010]

[22] Pelosato, R.; Cristiani, C.; Dotelli, G.; Mariani, M.; Donazzi, A.; Sora, I.N. Co-precipitation synthesis of SOFC electrode materials. *Int. J. Hydrogen Energy,* **2013**, *38*, 480-491. [http://dx.doi.org/10.1016/j.ijhydene.2012.09.063]

Ceramic Hollow Fibers: Fabrication and Application on Micro Solid Oxide Fuel Cells

Xiuxia Meng[1], Naitao Yang[1,*], Xiaoyao Tan[2] and Shaomin Liu[3,*]

[1] *School of Chemical Engineering, Shandong University of Technology, Zibo255049, China*

[2] *Department of Chemical Engineering, Tianjin Polytechnic University, Tianjin300387, China*

[3] *Department of Chemical Engineering, Curtin University, Perth, WA 6102, Australia*

Abstract: Micro-tubular solid oxide fuel cells (MT-SOFC) have recently attracted increasing attention due to their advantages, such as quick start-up/shut-down, high volumetric power density and better mobile/portable characteristics comparing favorably with planar and tubular SOFCs. Basically, there are four fabrication techniques to prepare MT-SOFCs: extrusion, cold isostatic pressing, slip-casting and phase-inversion. Among these techniques, the phase inversion method is a recently emerging technology to prepare the MT-SOFCs, which can be easily scaled up to realize an automatic and continuous preparation process. Therefore, this chapter gives a detailed account of the fabrication skills of micro-tubular SOFCs based on the phase-inversion techniques and the brief review of their performance test results.

Keywords: Ceramic membranes, Co-spinning, Dual-layer, Hollow fibre, Microtubular, Mobile, Phase inversion, Portable, Quick start-up/shut-down, Solid oxide fuel cells.

INTRODUCTION

Along with the energy crisis and global climate change, clean energy application and exploration of high-efficiency power sources have attracted more and more attention not only from the research communities but also from the industries [1, 2]. Among these efficient energy conversion devices, fuel cell is one typical example with the advantages of low emission and high energy efficiency. Fuel cells can directly convert chemical energy to electrical power *via* electrochemical

* **Corresponding authors Naitao Yang and Shaomin Liu:** School of Chemical Engineering, Shandong University of Technology, Zibo255049, China; Tel/Fax: +865332786292; E-mail: naitaoyang@126.com; Department of Chemical Engineering, Curtin University, Perth, WA 6102, Australia; Tel/Fax: +61-08-92669056; E-mail: shaomin.liu@curtin.edu.au

Moisés R. Cesário & Daniel A. de Macedo (Eds.)

redox. Since no gas turbine is involved, the fuel cell system is not limited by Carnot cycle, and therefore the efficiency of fuel cells are up to 80% or more if the waste-heat could be fully recycled, which attract intense research interest in the recent thirty years. Fuel cells have been classified into several types: alkaline fuel cell, proton exchange membrane fuel cell, phosphoric fuel cell, molten carbonate fuel cell, and solid oxide fuel cell. Among these various types, solid oxide fuel cells (SOFCs) are one of the most promising electrochemical devices featuring the advantages of higher energy efficiency, long term durability, noble-metal-free electrodes, anti-poisoning catalysts and high fuel choice flexibility including hydrogen, methane, CO, gasoline and other combustible gases. Therefore, SOFC has been considered as a promising power generator, transforming chemical energy into electricity, either in a power plant station or in a domestic electricity-heat generating unit. As the major part of SOFC components is fabricated from ceramic materials, such as ZrO_2 electrolyte membrane, perovskite cathode, cermet anode, ceramic bipolar connector and oxide sealant, SOFCs are also referred as ceramic fuel cells.

There are several configurations of SOFC design, including the planar, tubular (or microtubular), or honeycomb structure; among which planar and tubular geometries are the preferred designs [3 - 7]. The planar configuration is relatively an easier and cheaper design due to its simple fabrication protocols including tape casting to prepare ceramic membrane, screen printing to form thin layer electrode, and co-firing; these processing steps have been fully developed in the past decades and now are becoming established and mature techniques. And the efficiency of a planar SOFC has also been proved to be the highest compared to other designs due to the inherently shorter path for electron/ion transport from anode to cathode. However, this design has been found some challenges on thermal and mechanical stability. The shape distortion under high temperature is becoming more serious with a larger dimension of the single planar cell. Sealing is another issue which is challenging the high temperature operation because the sealing area is very large and complex associated with the planar design.

Apparently, the tubular SOFC configuration is a promising alternative due to the easier sealing process for its relative smaller sealing area. And the thermo-cycling behavior of a tubular cell has also been improved compared to the planar ones because of less shape distortion under high temperature operation. However, fuel cells with a tubular shape are more difficult to prepare than the planar structure. For example, the dense electrolyte layer on tubular ceramic support needs to be prepared by CVD-EVD method, or plasma spraying process [8, 9], which are more expensive and complex techniques than those employed in the planar fuel cells. Meanwhile, the energy efficiency, or power density in a tubular fuel cell is usually lower than the planar fuel cell. The reason is attributed to the long tubular structure, resulting in a long electronic path for current collection and leading to a higher ohmic loss.

MICRO TUBULAR SOLID OXIDE FUEL CELL: ADVANTAGES AND FABRICATIONS

Solid oxide fuel cells, either planar or tubular, are usually operated at the temperature of 750-900°C, which is too high to be practically applied, hindering their commercialization. High operation temperature will bring in many problems such as electrode sintering, chemical reaction between ceramic layers, corrosion of metal parts and finally damaging the life of the power generator. Furthermore, the start-up/set-down speed of the power source is an important threshold that limits the application of SOFC as an urgent power device or uninterruptible power supply (UPS). Large scale SOFCs are considered as promising power station due to their high efficiency of waste-heat and fuel recycling. On the other hand, small size of SOFC has been also emphasized, for its potential application as portable and mobile power unit [10].

Since 1990s, Kendall and co-workers have proposed an advanced cell design referred as the microtubular solid oxide fuel cell (MT-SOFC) [3]. It is not just an improvement to minimize the tubular structure of SOFC, but shows some remarkable advantages such as quick start-up/shut-down, high volumetric power density and better mobile/portable characteristics. When the diameter of the micro tubes reduces to 1-2 mm or less, its heat and mass transfer rates, and ion permeation rates are much quicker than that from large-thickness tubes. Therefore, new theoretical or modeling works are required to predict the performance of MT-SOFC. During the last two decades, a lot of research works have been done focusing on fabrication, theoretical analysis, modeling, and applications of MT-SOFCs. The advantages of MT-SOFC are re-emphasized as below:

High Volumetric Power Density. It is calculated that, the specific area of a micro tube with the diameter of 2 mm is 20 times larger than a tube with the diameter of 20 mm. Thus the volumetric power density of MT- SOFC is much higher than that of the large tubular one [11]. A cubic SOFC stack based on micro tubular configuration, designed by Suzuki *et al.* shows a power output up to 30W in a volume of 30 cm^3 (>1000 kW/m^3) [12]. This implies that small device with micro SOFC stacks can be applied to light mobile and portable power sources. In 2004, Nano Dynamics Inc., USA lunched a product

"Revolution50", based on MT-SOFC technology. This device could provide a power of 50W for 24h fueled with propane. It has potential market for infantries, outdoor adventurers, advertising boards or convenient battery chargers. Another application comes from the military attraction. Based on micro tubular SOFC, Acumentrics has deployed over 60,000 rugged UPS systems to the US military for use in forward-operating bases where performance under the worst environmental conditions is mission-critical. In the United Kingdom, a lightweight Skywalker aircraft has also been built by Adelan collaborating with Loughborough University, drove by micro tubular SOFCs.

Rapid Start-up/Shut-down Speed. MT-SOFC has very tiny chambers, which are beneficial for gas diffusion and heat transfer. The reacting gases are divided into trickle fluids by the micro channels and guided into the tiny chambers quickly. Both the mass and heat transfer are promoted due to the micro channels. A large-size SOFC stack usually faces the problem of heat distribution. Thermal stress will crack the ceramic electrolyte membrane and damage the whole cell stack. To decrease the temperature gradient, the large stacks must be heated up slowly, usually taking several hours. Similarly, the cooling speed of the large SOFC stack must be controlled at a very slow rate. In contrast, the start-up/shu--down of an MT-SOFC can be much faster because the micro channels can lead the heated gases to every corner rapidly, resulting to a much more uniform temperature distribution than that in the large-size SOFCs. For example, Kilbride operated a micro tubular SOFC with a heating/cooling speed up to 100 °C/min from 200 to 900°C for more than 50 circles, without any degradation caused by the thermal shock [13].

Lower Operation Temperature. Another obstacle of the commercialization of SOFCs is the sealing and aging problems. Decreasing the operation temperature seems to be an effective way to solve these problems. With currently accessible materials, feasible ways to decrease the operation temperature are to minimize the cell scale and to reduce the electrolyte thickness of the SOFCs. For example, Li and co-workers developed a micro tubular SOFC with $2.32W/cm^2$ at 600°C through a phase-inversion process [14]. As the operation is carried out at temperatures lower than 600°C, the sealing challenge has been much reduced and the iron-based alloys can be conveniently used as the connectors or covers. As a result, the structural and thermal stability of the SOFCs can be significantly improved and the cost of the power sources will be greatly reduced.

In general, the preparation methods of MT-SOFCs mainly include extrusion, cold isostatic pressing and phase-inversion spinning [15 - 19]. Extrusion method is developed based on plastic extrusion, which is a conventional way to prepare micro tubes. To prepare the mixture paste for extrusion, the ceramic powder is mixed with plastic materials with the addition of different binders (*e.g.* cellulose), pore formers (*e.g.* PMMA) and dispersants (*e.g.* water). Although the high-strength microtubular SOFC can be obtained *via* extrusion, the preparation method is time-consuming and often very expensive. The resultant substrate layers with low porosities often have high diffusion resistance. It is also hard to obtain multi-layer micro tube *via* one extrusion step. Therefore, the cell components have to be made by a multi-step method *via* a layer-by-layer process, leading to a time-consuming and complex preparation.

Cold isostatic pressing is another method to prepare micro ceramic tubes. Since the cold isostatic pressing often offers a high and well-distributed pressure for products, the resulted fuel cells by this method possess a high structural density and high mechanical strength; therefore it is more suitable for the fabrication of dense electrolyte micro tubes than porous electrode with pore formers being applied. The performance of the cells based on the cold isostatic pressing is often low because of the prepared electrode micro tubes with low porosities hindering the gas diffusion or thick electrolyte layer with large ohmic resistance.

In comparison with extrusion and cold isostatic pressing, the phase inversion is an interesting method to prepare the micro tubes (also referred as ceramic hollow fibers). The prepared micro tubes can be designed to be in highly asymmetric structure with ordered porosities, which facilitate the gas transfer through the substrate layer and improve the overall output of MT-SOFCs. Therefore, the main concern of this chapter has been placed on the development of MT-SOFCs based on phase inversion technique and we highlight the recent significant progress made on this subject.

BASIC PRINCIPLES OF THE PHASE INVERSION TECHNIQUE

The phase inversion was firstly introduced to prepare the asymmetric polymeric membranes in 1960s. The ceramic micro tubes have been established with the combing process of phase inversion and sintering technique since 1999 by Li's group [20 - 22]. Now, the technique has been widely used in the preparation of inorganic hollow fibers. The basic process generally includes three steps as shown in Fig. (**1**). Firstly, a spinning suspension is prepared, consisting of polymer binder, solvent, additive and inorganic powders. The polymer binder and the additives are added into the solvent(s), and completely dissolved under strongly stirring conditions to form homogenous polymer binder solution. Then the pretreated inorganic powders are introduced into the solution and uniformly dispersed for 24 hours under stirring condition to form a uniform spinning suspension (also referred as the dope). The dope is degassed under vacuum to remove the gas bubbles trapped inside the suspension. Secondly, the hollow fiber precursors are span *via* a spinneret. The degassed suspension is pressurized through a tube-in-orifice spinneret into a coagulation bath to solidify and form solid hollow fiber precursors. The phase inversion process, during which the polymer is transferred from liquid state to solid, is immediately occurring once the nascent hollow fiber precursor is in contact with the coagulants. The different precipitation rate of polymer inside the hollow fiber precursor leads to the asymmetric structure consisting of porous finger-like pores and relatively dense sponge-like layer of the resultant hollow fibers with one typical membrane morphology in Fig. (**2**). Thirdly, the high-temperature sintering is performed to remove the polymer binder and other organic additives, and to bond the inorganic particles into ceramic product with certain mechanical strength. After the sintering process, the overall morphology from the precursor fiber is generally preserved but the microstructure is significantly changed. For example, the sponge-like voids become denser. Because the relatively dense layer and porous finger-like pores are formed in one step, simplifying the preparation procedure.

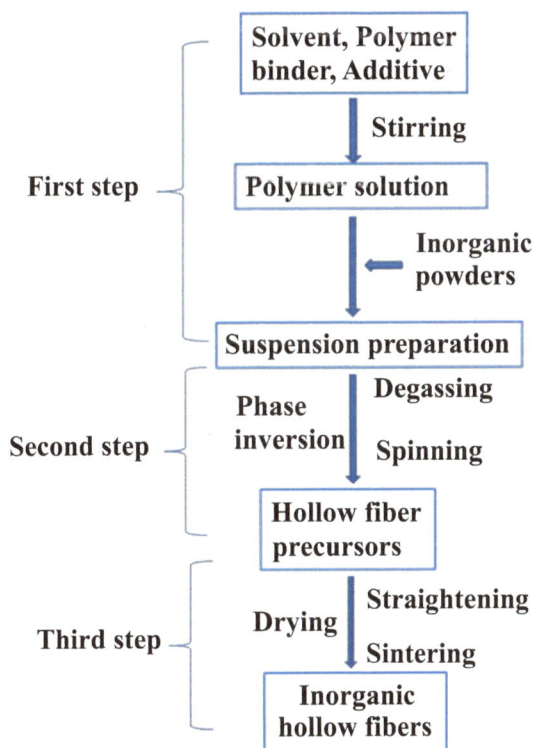

Fig. (1). The preparation procedure of inorganic hollow fiber based on phase inversion technique.

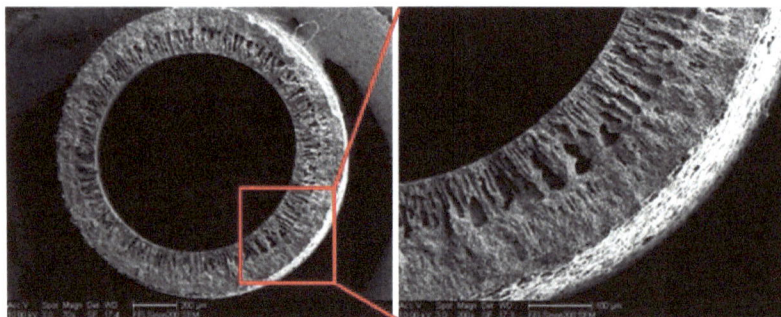

Fig. (2). SEM images of the asymmetric structure of hollow fibre precursors based on phase inversion technique.

SINGLE-LAYER SPINNING FOR SOFC

Single-layer spinning is a well-developed method initially applied for the preparation of organic hollow fiber membranes [23, 24]. Fig. (**3**) shows the schematic diagram of the single-layer spinning to prepare hollow fibers. When the required amount of ceramic powder is mixed inside the polymer solution, the ceramic hollow fiber precursor can be extruded in a similar manner with the polymeric fibers; thus this technique has been modified to prepare ceramic hollow fiber membranes with subsequent sintering at high temperatures [25]. Fig. (**4**) shows several typical ceramic hollow fibers prepared *via* the phase-inversion method in our laboratory, such as Ni-YSZ (yttria stabilized Zirconia) anode [17], LSCF ($La_{0.6}Sr_{0.4}Co_{0.2}Fe_{0.8}O_{3-\delta}$) cathode [26] and YSZ electrolyte [27]. As the spinning is carried out at room temperature, no expensive apparatus is required and thus the fabrication process is very straightforward and inexpensive.

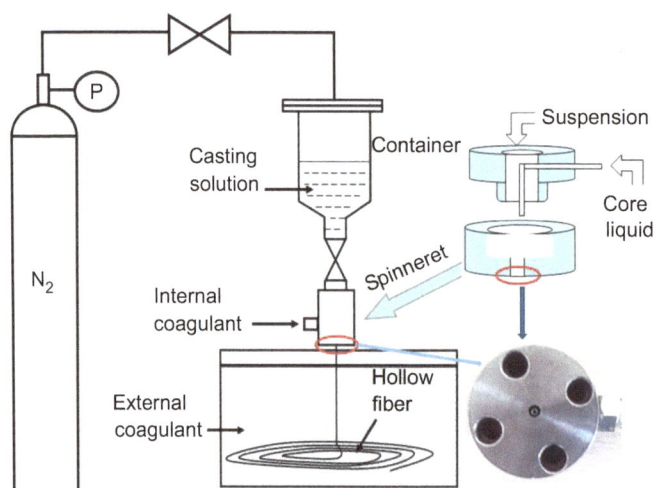

Fig. (3). Setup for spinning hollow fiber membranes.

Based on the electrolyte hollow fibers, the MT-SOFCs have been developed and SDC ($Sm_{0.2}Ce_{0.8}O_{1.9}$) electrolyte-supported MT-SOFCs are prepared, shown in Fig. (**5**). These MT-SOFCs exhibit high mechanical and thermal prosperities. However, the output of these MT-SOFCs is still low due to the thick electrolyte layer, which induces a high ohmic resistance.

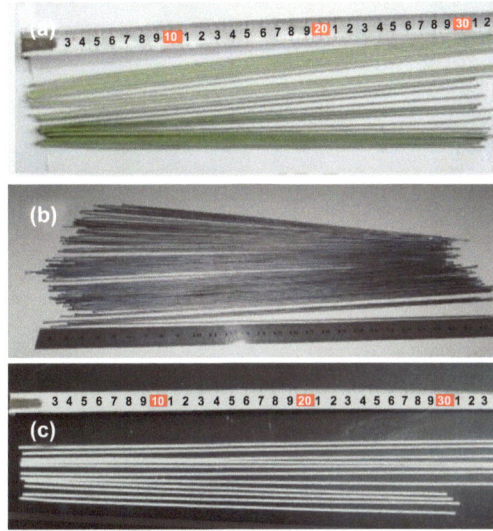

Fig. (4). Photos of some electrolyte ceramic hollow fibers prepared by the phase-inversion method (**a**) Ni-YSZ [17], (**b**) LSCF [26] and (**c**) YSZ [27].

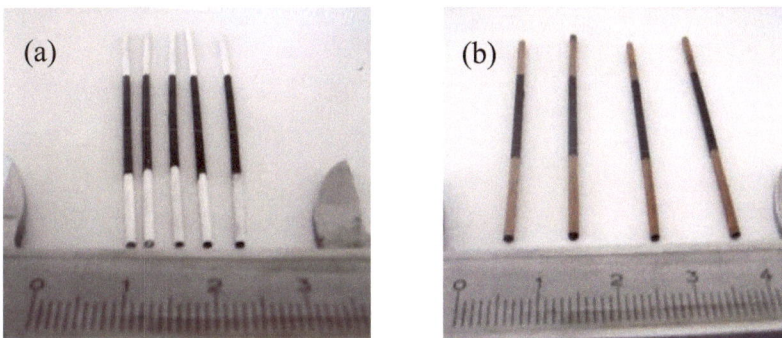

Fig. (5). Photos of MT-SOFCs based on electrolyte hollow fibers of YSZ (**a**) and SDC (**b**).

To improve the performance and decrease the electrolyte thickness, anode-supported MT-SOFCs are prepared in light of the following process shown in Fig. (**6**). For instance, an anode hollow fiber should be prepared firstly by phase-inversion process with photos shown in Fig. (**4c**). After high temperature sintering, it can be functioned as an anode supporter and anode current collector. Electrolyte powder such as YSZ should be dispersed inside the ethanol solvent with the additive such as binder, surfactant and plasticizer together to form a stable suspension. The anode hollow fiber is firstly immersed inside the electrolyte suspension and pulled out with a uniform speed. After a few repetitions, a thin layer of YSZ membrane is formed; subsequently, the coated YSZ electrolyte film and anode support are co-sintered at 1400°C to form a dual-layer half-cell. After this, cathode layer is coated and fired in similar process. The cathode suspension can be prepared with LSM perovskite powder using a similar method as the YSZ suspension. At the end, the thickness of the prepared dense electrolyte membrane and porous LSM layer is around 10 and 20 µm, respectively. The power density of the micro tubular cell based on a Ni-YSZ hollow fiber reaches up to 0.8Wcm^{-2} at 850°C, using H_2 as fuel and air as oxidant [17, 24]. However, the cells show weak stability after a few thermal cycles, because the entire cell is based on the Ni-YSZ anode supported hollow fiber, and the other layers are prepared separately by coating method with layer by layer. The different layers are not well integrated with the other leaving many defects or voids in the interfaces or in the boundary area. When the cell is operated under thermal cycles, these layers are easily delaminated or peeled off. Therefore, this multi-step process combining single-layer spinning and layer-by-layer coating method to prepare hollow fiber ceramic fuel cells is time-consuming and suffering weak stability. Thus, to overcome these problems, we are inspired to develop the hollow fiber SOFCs with multi-layers in one spinning process.

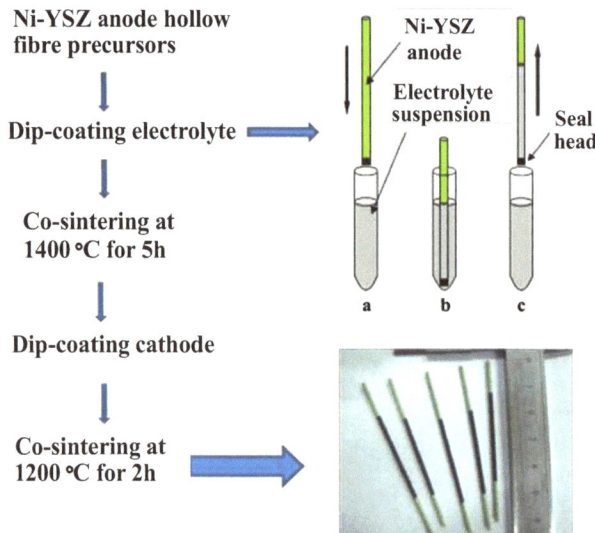

Fig. (6). Preparation process of MT-SOFCs based on the anode supports [17, 22].

To increase the stability against the thermal shock without the delaminating problem, a novel MT-SOFC with an integrated electrolyte/anode structure has been developed in our group based on the highly asymmetric structure of the electrolyte hollow fibers [28 - 30]. The microstructure of the hollow fiber membranes can be well-tailored by controlling the suspension composition and the spinning processing parameters. For example, when the composition of internal coagulate is controlled at 30wt.%ethanol/70wt.%NMP, the resultant YSZ hollow fibers exhibit a desired asymmetric structure consisting of a thin dense skin layer and a thick porous substrate layer with the long finger-like pores and honeycomb inner surface (Fig.7). After vacuum-assistant impregnation of anode catalyst Ni in the porous structure, an integrated anode/electrolyte structure can be formed. Porous YSZ impregnated with Ni particles served as anode support layer and the thin dense skin layer can directly serve as the electrolyte membrane. Because the thin dense skin layer and the porous substrate are formed in one step from the same material, the electrolyte membrane can be perfectly integrated with the anode. The process has been successfully employed to fabricate anode-supported Ni-YSZ/YSZ/LSCF and Ni-YSZ/YSZ/Ag MT-SOFCs. The maximum output of the resultant fuel cells with the configuration of Ni-YSZ/YSZ/Ag reached up to 576 mWcm^{-2} at 800 °C with H_2 as the fuel feed and air as the oxidant [29]. Furthermore, the cells exhibit high stable output in fifty quick thermal cycles because the catalyst is not part of the mechanical backbone. Noteworthy is that the output of MT-SOFCs is strongly related to the microstructure of the micro tube. However, it is still a great challenge to precisely design and control the macro- and microstructure of the inorganic hollow fiber membranes because too many factors pose their influence. These factors include the humidity of the laboratory atmosphere, ceramic particle size and its distribution, the shape and the surface property of inorganic powders, the composition and viscosity of the spinning suspension, the spinning conditions (spinning rate, air gap, internal coagulant) and the sintering parameters (sintering temperature, dwelling time, heating rate) can solely or jointly affect the formation of membrane structures.

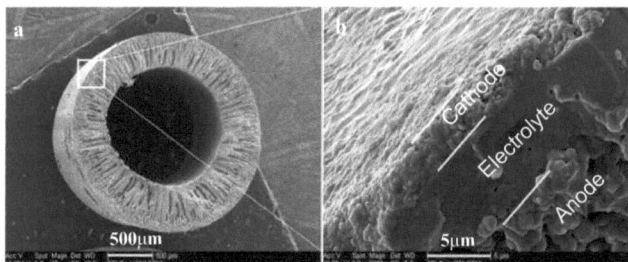

Fig. 7 contd.....

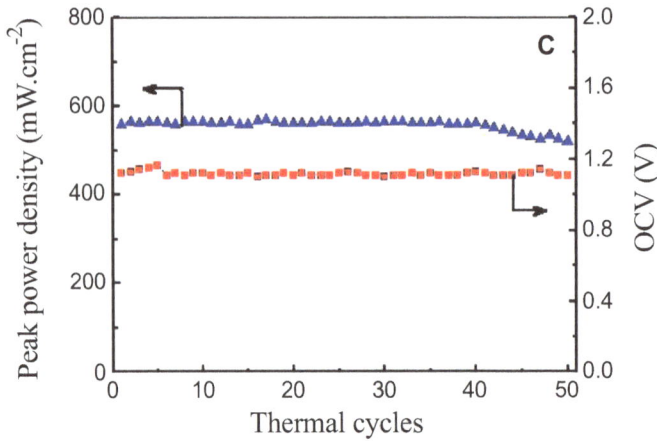

Fig. (7). Microstructure of a prepared micro tubular SOFC and performance variation with thermal cycles. (**a**) cross section; (**b**) skin area with cathode/electrolyte/anode; (**c**) OCVs and peak power densities under thermal cycles.

DUAL-LAYER CO-SPINNING FOR MT-SOFC

As discussed in the previous section, different components of the SOFC have to be prepared separately *via* multi-step method on these single-layered hollow fibers, which is time-consuming and introduces many defects. Obviously, multilayer spinning has been considered as a great advancement to fabricate these hollow fiber SOFCs compared to the single-layer spinning. The phase inversion method has been then modified to suit a dual-layer co-spinning process [14, 31 - 35]. As shown in Fig. (**8**), in this dual-layer spinning process, a newly designed spinneret with two orifices is adopted. Two different dopes are pressurized simultaneously through the two orifices (Fig. **9**) into the coagulant bucket and thus dual-layer micro tubes will be obtained. During the initial stage of the spinning, parts of the two spinning dopes may penetrate to each other to some extent at the phase boundary area prior to submerging into the water (external coagulant) leading to mutual adhesion avoiding the crack formation between two layers after co-sintering. Li's group has prepared a GDC($Gd_{0.2}Ce_{0.8}O_{1.9}$)/Ni-GDC dual-layer hollow fiber using this co-spinning method and they perform the coating of the cathode layer on the prepared GDC/Ni-GDC dual-layer hollow fibers to construct the integrated MT-SOFCs with full components [14]. After continuous improvement such as reducing the electrolyte layer thickness, tailoring the microstructure of anode, the high-performance MT-SOFCs have been achieved. For example, the power density of the fuel cells with the configuration of LSCF/GDC/Ni-GDC reached up to 2.32 W/cm^2 at 600 °C when the thickness of the electrolyte is 10 μm and the anode is of 70% finger-like voids with reduced gas diffusion resistance [14]. However, it should be noted that the cell is mechanically weak due to the inherent fragility of CeO_2-based ceramic materials.

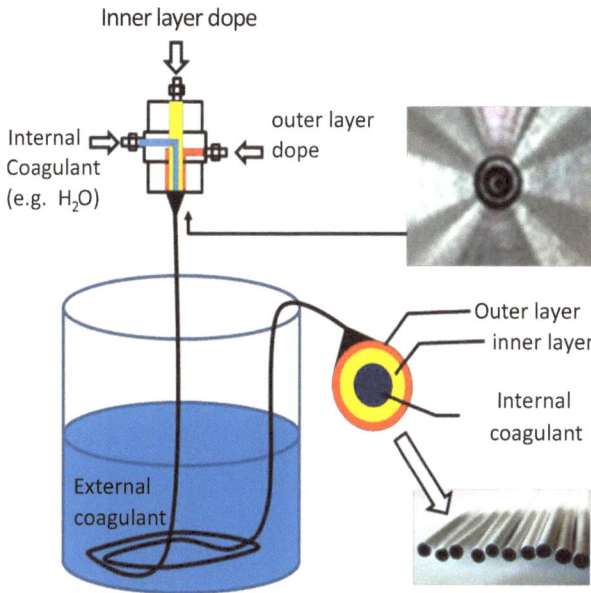

Fig. (8). Schematic diagram for spinning two-layered ceramic hollow fibers [36].

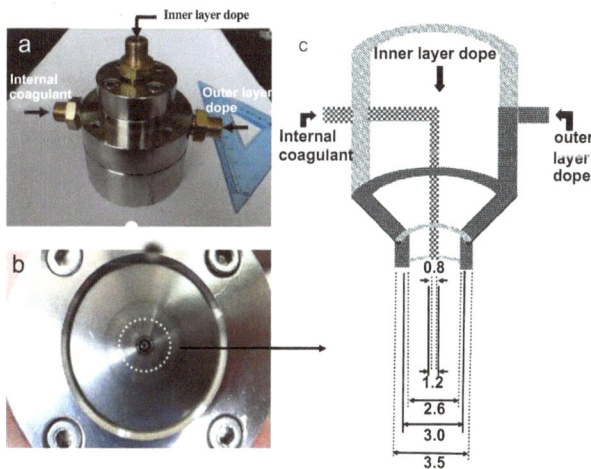

Fig. (9). Photographic images of triple orifice spinneret from (**a**) side view, (**b**) bottom, and (**c**) sectional design and dimension [35].

Considered that the hollow fiber SOFCs are so tiny and thin, more robust ceramics like zirconia based ceramics are proposed as the electrolyte material to fabricate the micro SOFCs. Following the pioneering work by Li and co-workers, our group also explored the dual-layer spinning method to prepare MT-SOFC using conventional electrolyte of YSZ [31, 34 - 36]. The cathode-supported MT-

SOFCs have been successfully prepared based on YSZ/LSM(La$_{0.8}$Sr$_{0.2}$MnO$_{3-\delta}$)-YSZ dual-layer hollow fiber after coating Ni-YSZ anode [34]. With the increase of sintering temperature, the mechanical strength of the hollow fiber and compactness of electrolyte with cathode is improved while the porosity of the cathode layer decreases. To obtain the applicable YSZ/LSM-YSZ dual-layer hollow fiber appropriate for MT-SOFCs, the sintering temperature should be controlled between 1350 and 1400°C. However, the resultant MT-SOFCs exhibited lower peak power density of 290 mWcm^{-2} at 850°C using the hydrogen as the fuel and air as oxidant than that reported from the literature. The reason for this undesired performance is due to the high sintering temperature at which, the sintered cathode layer is too densified with increased cathode polarization resistance. To further improve the performance of cathode-supported MT-SOFCs, the cathode functional layer is inserted between cathode and electrolyte, and therefore the LSM-YSZ/LSM dual-layer hollow fibers is applied as the substrate to support the MT-SOFCs (Fig. **10**) [37]. The performance of the resultant cathode-supported MT-SOFCs has been improved to 475 mWcm^{-2} but the polarization resistance still limited the output of the cells due to high sintering temperature of cathode.

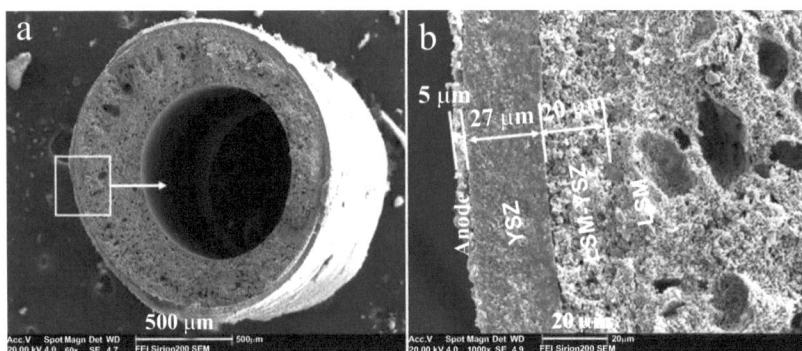

Fig. (10). Microstructure of dual-layer cathode-supported MT-SOFC. (**a**) cross-section; (**b**) anode/electrolyte/ LSM-YSZ40/LSM interface [37].

Comparing with the cathode-supported MT-SOFC, the anode-supported cell has been improved due to their high output and good mechanical strength. Based on the co-spinning/co-sintering process, YSZ/Ni-YSZ dual-layer hollow fibers were prepared and applied for MT-SOFC fabrication [35, 38]. The electrolyte thickness can be controlled by varying the spinning rate from 0.5 to 2 ml min^{-1}. Meanwhile, the anode structure is tailored through the composition of anode dope due to the addition of non-solvent ethanol leading to the variation of the viscosity. The obtained YSZ/Ni-YSZ dual-layer hollow fibers possess asymmetric microstructure with dense electrolyte (thickness of 12 µm) and porous anode layer (porosity 35%). After coating the LSM cathode, the peak power density of the fuel cell is 486 mW/cm^2. However, the anode-supported MT-SOFCs are not stable using the hydrocarbon as fuel due to the carbon deposition on the Ni catalyst. Particularly, a Ni-YSZ/porous YSZ dual-layer hollow fiber is designed as the base of a carbon-resistant MT-SOFC, where Cu-CeO$_2$catalyst is deposited in the porous YSZ layer to form a carbon-resistant anode and the Ni-YSZ layer is optimized to be the functional anode layer (Fig. **11**) [31]. The resultant fuel cell exhibits a peak power density of 566 mW/cm^2 at 850 °C with methane feed and has maintained a stable output for more than 120 hours, as shown in Fig. (**12**).

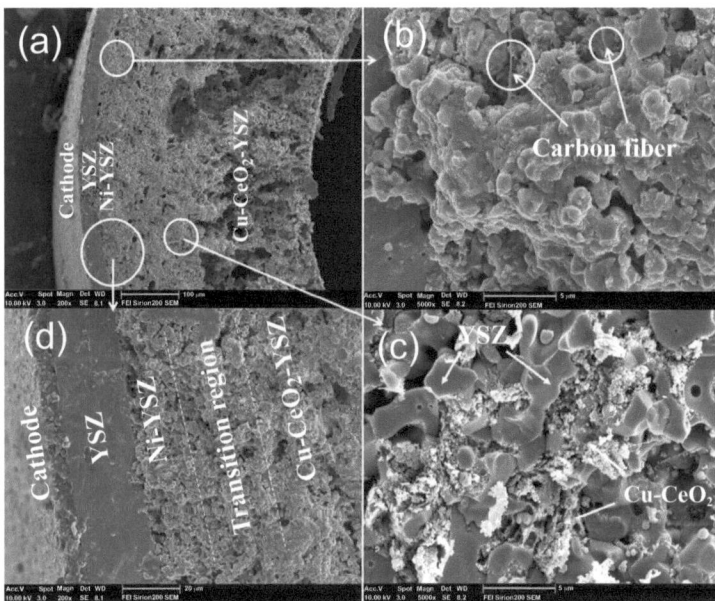

Fig. (11). SEM images of carbon morphology of a direct-methane MT-SOFC after long-term test [31]. (**a**) cross-section, (**b**) Ni-YSZ layer (anode-1), (**c**) Cu-CeO$_2$-YSZ layer, (**d**) the transition region between Ni-YSZ layer and Cu-CeO$_2$-YSZ.

TRIPLE-LAYER CO-SPINNING

Recently, triple-layer co-spinning has been reported by Li's group based on the same phase inversion technique [39, 40]. The CGO/40%NiO-60%CGO/ 60%NiO-40%CGO triple-layer hollow fiber (Fig. **13b**) [39, 40] was successfully fabricated in one step using a four-orifice spinneret (Fig. **13a**), in which the interlayer is thought to be anode functional layer (AFL). By adjusting the spinning rate of AFL during co-spinning, different AFL thickness is obtained contributing a great buffer layer for a good integration of the two neighboring layers. The gas-tightness of electrolyte layer and mechanical strength are remarkably improved due to the introduction of AFL. Unfortunately, the existence of finger-like pores inside electrolyte layer is unfavorable to maintain the gas-tightness of the electrolyte. Much more future works are still needed to optimize the microstructure of the MT-SOFCs *via* the triple-layer co-spinning and sintering technique. Some suggestions are proposed to eliminate the large voids in the electrolyte layer, such as increasing viscosity of spinning suspension, using a higher air gap or increasing the spinning rate of internal coagulant.

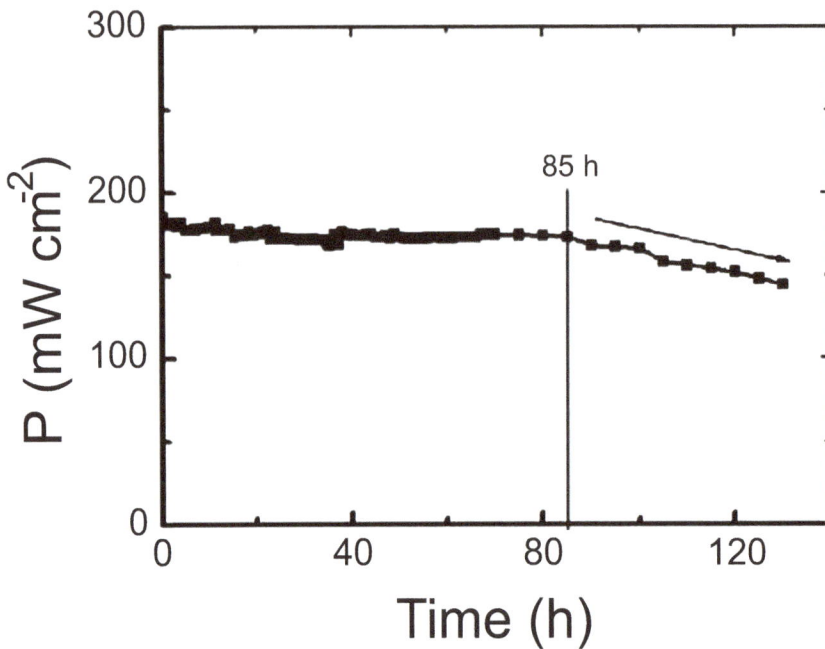

Fig. (12). Power density at 0.7V as a function of time for MT-SOFC supported by the Ni-YSZ/Cu-CeO$_2$-YSZ hollow fibre anode (anode-1) operating on methane at 800 °C [31].

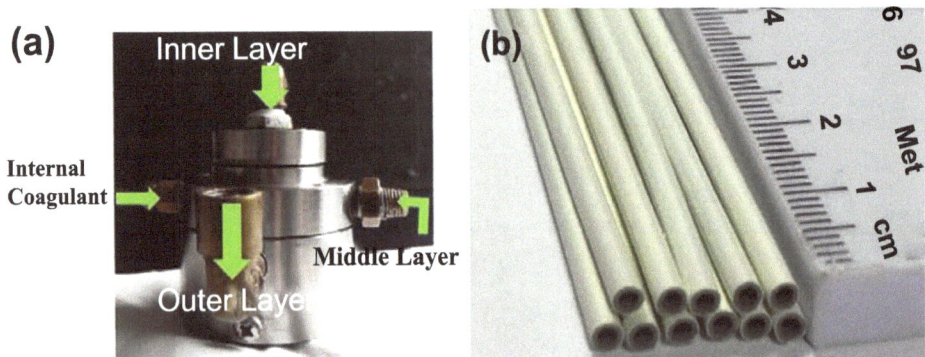

Fig. (13). Photographic pictures of (**a**) quadruple-orifice spinneret; (**b**) example of triple-layer precursors [39, 40].

CONCLUSION REMARKS

In this chapter, the potential of micro-tubular SOFCs and their preparation route have been reviewed with more emphasis on the phase inversion and sintering techniques to provide the ceramic micro tubes. As a kind of electrochemical device, SOFCs are consisted by many components thus their fabrication or assembling is completed by multi-steps *via* layer by layer technique; this is undoubtedly time-consuming and full of uncertainties. This is also applied to the synthesis of MT-SOFCs. To simply the preparation process and increase the stability of the cell, the phase inversion method has been modified and advanced to prepare various micro ceramic tubes from single-layer to dual-layer or triple-layer. Despite of significant progress, the challenges are still there, which need more work to further optimize the preparation conditions for better energy output and operational stability. It can be optimistically estimated that the MT-SOFC is still on its embryonic stage but it will continue to attract the intensive interest from researchers until it realizes its commercialization. During this process, ceramic hollow fiber technology will play a key role. Hopefully, this book chapter provides a straightforward guideline for beginners to take up this to further optimize the MT-SOFC system and realize its potential for sustainable and clean energy development.

CONFLICT OF INTEREST

The authors confirm that they have no conflict of interest to declare for this publication.

ACKNOWLEDGEMENTS

The authors acknowledge the research support by the Natural Science Foundation of China (21376143, 21176187) and the Australian Research Council through the Future Fellow Program (FT12100178).

REFERENCES

[1]   IEA. Solutions for the 21st century-zero emissions technologies for fossil fuels: technology status report. In: *International energy agency, committee on energy research and technology, working party on fossil fuels*; , **2002**; 12, pp. 3-32.

[2]   La, O.G.; In, H.J.; Crumlin, E.; Barbastathis, G.; Shao-Horn, Y. Recent advances in microdevices for electrochemical energy conversion and storage. *Int. J. Energy Res.,* **2007**, *31*, 548-575. [http://dx.doi.org/10.1002/er.1280]

[3]   Kendall, K. Tubular Cells. *A study of a tubular solid oxide fuel cell [C]. Final Year Project Report.,* **1993**, 1-86. Middlesex University.

[4]   Kendall, K. Progress in Microtubular Solid Oxide Fuel Cells. *Int J Appl Ceram Tech,* **2010**, *7*, 1-9.
[http://dx.doi.org/10.1111/j.1744-7402.2008.02350.x]

[5]   Lawlor, V. Review of the micro-tubular solid oxide fuel cell (Part II: Cell design issues and research activities). *J. Power Sources,* **2013**, *240*, 421-441.
[http://dx.doi.org/10.1016/j.jpowsour.2013.03.191]

[6]   Minh, N.Q. Ceramic Fuel Cells. *J. Am. Ceram. Soc.,* **1993**, *76*, 563-588.
[http://dx.doi.org/10.1111/j.1151-2916.1993.tb03645.x]

[7]   Ormerod, R.M. Solid oxide fuel cells. *Chem. Soc. Rev.,* **2003**, *32*(1), 17-28.
[http://dx.doi.org/10.1039/b105764m] [PMID: 12596542]

[8]   Huang, K.; Singhal, S.C. Cathode-supported tubular solid oxide fuel cell technology: A critical review. *J. Power Sources,* **2013**, *237*, 84-97.
[http://dx.doi.org/10.1016/j.jpowsour.2013.03.001]

[9]   Yan, J.; Dong, Y.; Yu, C.; Jiang, Y. Fabrication and characterization of anode substrates and supported electrolyte thin films for intermediate temperature solid oxide fuel cells. *J Inorg Mater,* **2001**, *16*, 804-814.

[10]   Kendall, K.; Meadowcroft, A. Improved ceramics leading to microtubular Solid Oxide Fuel Cells (mSOFCs). *Int. J. Hydrogen Energy,* **2013**, *38*, 1725-1730.
[http://dx.doi.org/10.1016/j.ijhydene.2012.08.094]

[11]   Suzuki, T.; Yamaguchi, T.; Fujishiro, Y.; Awano, M. Current collecting efficiency of micro tubular SOFCs. *J. Power Sources,* **2007**, *163*, 737-742.
[http://dx.doi.org/10.1016/j.jpowsour.2006.09.071]

[12]   Suzuki, T.; Funahashi, Y.; Yamaguchi, T.; Fujishiro, Y.; Awano, M. Development of cube-type SOFC stacks using anode-supported tubular cells. *J. Power Sources,* **2008**, *175*, 68-74.
[http://dx.doi.org/10.1016/j.jpowsour.2007.09.082]

[13]   Kilbride, I.P. Preparation and properties of small diameter tubular solid oxide fuel cells for rapid start-up. *J. Power Sources,* **1996**, *161*, 167-171.
[http://dx.doi.org/10.1016/S0378-7753(96)02362-2]

[14]   Othman, M.H.; Droushiotis, N.; Wu, Z.; Kelsall, G.; Li, K. High-performance, anode-supported, microtubular SOFC prepared from single-step-fabricated, dual-layer hollow fibers. *Adv. Mater.,* **2011**, *23*(21), 2480-2483.
[http://dx.doi.org/10.1002/adma.201100194] [PMID: 21484892]

[15]   Suzuki, T.; Yamaguchi, T.; Fujishiro, Y.; Awano, M. Improvement of SOFC Performance Using a Microtubular, Anode-Supported SOFC. *J. Electrochem. Soc.,* **2006**, *153*, A925-A8.
[http://dx.doi.org/10.1149/1.2185284]

[16]   Campana, R.; Merino, R.I.; Larrea, A.; Villarreal, I.; Orera, V.M. Fabrication, electrochemical characterization and thermal cycling of anode supported microtubular solid oxide fuel cells. *J. Power Sources,* **2009**, *192*, 120-125.
[http://dx.doi.org/10.1016/j.jpowsour.2008.12.107]

[17]   Yang, N.; Tan, X.; Ma, Z. A phase inversion/sintering process to fabricate nickel/yttria-stabilized zirconia hollow fibers as the anode support for micro-tubular solid oxide fuel cells. *J. Power Sources,* **2008**, *183*, 14-19.
[http://dx.doi.org/10.1016/j.jpowsour.2008.05.006]

[18]   Yang, N.; Tan, X.; Ma, Z.; Thursfield, A. Fabrication and Characterization of $Ce_{0.8}Sm_{0.2}O_{1.9}$ Microtubular Dual-Structured Electrolyte Membranes for Application in Solid Oxide Fuel Cell Technology. *J. Am. Ceram. Soc.,* **2009**, *92*, 2544-2550.
[http://dx.doi.org/10.1111/j.1551-2916.2009.03267.x]

[19]   Jamil, S.M.; Othman, M.H.; Rahman, M.A.; Jaafar, J.; Ismail, A.F.; Li, K. Recent fabrication techniques for micro-tubular solid oxide fuel cell support: A review. *J. Eur. Ceram. Soc.,* **2015**, *35*, 1-22.
       [http://dx.doi.org/10.1016/j.jeurceramsoc.2014.08.034]

[20]   Liu, S.; Li, K.; Hughes, R. Preparation of porous aluminium oxide (Al_2O_3) hollow fibre membranes by a combined phase-inversion and sintering method. *Ceram. Int.,* **2003**, *29*, 875-881.
       [http://dx.doi.org/10.1016/S0272-8842(03)00030-0]

[21]   Tan, X.; Liu, S.; Li, K. Preparation and characterization of inorganic hollow fiber membranes. *J. Membr. Sci.,* **2001**, *188*, 87-95.
       [http://dx.doi.org/10.1016/S0376-7388(01)00369-6]

[22]   Tan, X.; Li, K. Porous Membrane Reactors. In: *Inorganic Membrane Reactors*; John Wiley & Sons, Ltd: USA, **2014**; pp. 27-73.
       [http://dx.doi.org/10.1002/9781118672839.ch2]

[23]   Liu, Y.; Chen, O.Y.; Wei, C.C.; Li, K. Preparation of yttria-stabilised zirconia (YSZ) hollow fibre membranes. *Desalination,* **2006**, *199*, 360-362.
       [http://dx.doi.org/10.1016/j.desal.2006.03.216]

[24]   Yang, N.T.; Tan, X.; Meng, X.X.; Yin, Y.; Ma, Z-F. Fabrication of anode supported micro tubular SOFCs by dip-coating process on NiO/YSZ hollow fibers. *216th ECS Meeting.,* **2009**ECS TransactionVienna, Austria, pp. 811-888.

[25]   Tan, X.; Liu, Y.; Li, K. Mixed conducting ceramic hollow-fiber membranes for air separation. *AIChE J.,* **2005**, *51*, 1991-2000.
       [http://dx.doi.org/10.1002/aic.10475]

[26]   Tan, X.; Liu, N.; Meng, B.; Sunarso, J.; Zhang, K.; Liu, S. Oxygen permeation behavior of $La_{0.6}Sr_{0.4}Co_{0.8}Fe_{0.2}O_3$ hollow fibre membranes with highly concentrated CO_2 exposure. *J. Membr. Sci.,* **2012**, *389*, 216-222.
       [http://dx.doi.org/10.1016/j.memsci.2011.10.032]

[27]   Liu, L.; Tan, X.; Liu, S. Yttria Stabilized Zirconia Hollow Fiber Membranes. *J. Am. Ceram. Soc.,* **2006**, *89*, 1156-1159.
       [http://dx.doi.org/10.1111/j.1551-2916.2005.00901.x]

[28]   Yin, W.; Meng, B.; Meng, X.; Tan, X. Highly asymmetric yttria stabilized zirconia hollow fibre membranes. *J. Alloys Compd.,* **2009**, *476*, 566-570.
       [http://dx.doi.org/10.1016/j.jallcom.2008.09.079]

[29]   Liu, Y.; Liu, N.; Tan, X. Preparation of microtubular solid oxide fuel cells based on highly asymmetric structured electrolyte hollow fibers. *Sci. China Chem.,* **2011**, *54*, 850-855.
       [http://dx.doi.org/10.1007/s11426-010-4220-8]

[30]   Meng, X.; Yan, W.; Yang, N.; Tan, X.; Liu, S. Highly stable microtubular solid oxide fuel cells based on integrated electrolyte/anode hollow fibers. *J. Power Sources,* **2015**, *275*, 362-369.
       [http://dx.doi.org/10.1016/j.jpowsour.2014.11.027]

[31]   Meng, X.; Gong, X.; Yin, Y.; Yang, N.; Tan, X.; Ma, Z-F. Effect of the co-spun anode functional layer on the performance of the direct-methane microtubular solid oxide fuel cells. *J. Power Sources,* **2014**, *247*, 587-593.
       [http://dx.doi.org/10.1016/j.jpowsour.2013.08.133]

[32]   Othman, M.H.; Droushiotis, N.; Wu, Z.; Kanawka, K.; Kelsall, G.; Li, K. Electrolyte thickness control and its effect on electrolyte/anode dual-layer hollow fibres for micro-tubular solid oxide fuel cells. *J. Membr. Sci.,* **2010**, *365*, 382-388.
       [http://dx.doi.org/10.1016/j.memsci.2010.09.036]

[33] Othman, M.H.; Wu, Z.; Droushiotis, N.; Doraswami, U.; Kelsall, G.; Li, K. Single-step fabrication and characterisations of electrolyte/anode dual-layer hollow fibres for micro-tubular solid oxide fuel cells. *J. Membr. Sci.,* **2010**, *351*, 196-204.
[http://dx.doi.org/10.1016/j.memsci.2010.01.050]

[34] Meng, X.; Gong, X.; Yang, N.; Tan, X.; Yin, Y.; Ma, Z-F. Fabrication of Y_2O_3-stabilized-ZrO_2(YSZ)/$La_{0.8}Sr_{0.2}MnO_{3-\alpha}$–YSZ dual-layer hollow fibers for the cathode-supported micro-tubular solid oxide fuel cells by a co-spinning/co-sintering technique. *J. Power Sources,* **2013**, *237*, 277-284.
[http://dx.doi.org/10.1016/j.jpowsour.2013.03.026]

[35] Meng, X.; Gong, X.; Yin, Y.; Yang, N-T.; Tan, X.; Ma, Z-F. Microstructure tailoring of YSZ/Ni-YSZ dual-layer hollow fibers for micro-tubular solid oxide fuel cell application. *Int. J. Hydrogen Energy,* **2013**, *38*, 6780-6788.
[http://dx.doi.org/10.1016/j.ijhydene.2013.03.088]

[36] Meng, X.; Gong, X.; Yang, N.; Tan, X.; Yin, Y.; Ma, Z-F. Carbon-resistant Micro Tubular SOFCs Fabricated by Co-spinning Process Based on a Phase-inversion Method. *ECS Trans.,* **2013**, *57*, 1259-1266.
[http://dx.doi.org/10.1149/05701.1259ecst]

[37] Meng, X.; Yang, N.; Gong, X.; Yin, Y.; Ma, Z-F.; Tan, X. Novel cathode-supported hollow fibers for light weight micro-tubular solid oxide fuel cells with an active cathode functional layer. *J. Mater. Chem. A Mater. Energy Sustain.,* **2015**, *3*, 1017-1022.
[http://dx.doi.org/10.1039/C4TA04635H]

[38] Gong, X.; Meng, X-X.; Yang, N-T.; Tan, X-Y.; Yin, Y-M.; Ma, Z-F. Electrolyte Thickness Control and Its Effect on YSZ/Ni-YSZ Dual-layer Hollow Fibres. *J Inorg Mater,* **2013**, *28*, 1108-1114.
[http://dx.doi.org/10.3724/SP.J.1077.2013.12755]

[39] Li, T.; Wu, Z.; Li, K. Single-step fabrication and characterisations of triple-layer ceramic hollow fibres for micro-tubular solid oxide fuel cells (SOFCs). *J. Membr. Sci.,* **2014**, *449*, 1-8.
[http://dx.doi.org/10.1016/j.memsci.2013.08.009]

[40] Li, T.; Wu, Z.; Li, K. Co-extrusion of electrolyte/anode functional layer/anode triple-layer ceramic hollow fibres for micro-tubular solid oxide fuel cells–electrochemical performance study. *J. Power Sources,* **2015**, *273*, 999-1005.
[http://dx.doi.org/10.1016/j.jpowsour.2014.10.004]

CHAPTER 7

Electrolyte Hollow Fiber as Support *via* Phase-Inversion-Based Extrusion/Sintering Technique for Micro Tubular Solid Oxide Fuel Cell

Mohd Hafiz Dzarfan Othman[*], **Siti Munira Jamil**, **Mukhlis A. Rahman**, **Juhana Jaafar** and **A.F. Ismail**

Advanced Membrane Technology Research Centre, Universiti Teknologi Malaysia, 81310 UTM Johor Bahru, Johor, Malaysia

Abstract: This chapter focuses to discuss about the recently introduced phase-inversion based extrusion technique for fabrication of electrolyte hollow fiber support and multi-layer of the micro tubular solid oxide fuel cell (MT-SOFC). The effects of different fabrication parameters on the morphologies and electrochemical performances of developed electrolytes *via* the technique were critically discussed. The future direction of this advanced phase-inversion-based extrusion technique in the MT-SOFCs fabrication was also being discussed at the end of this book chapter.

Keywords: Anode-supported, Co-extrusion, Co-sintering, Electrolyte-supported, Hollow fiber, Multi-layer, Phase-inversion, Single-layer, SOFC, Thin layer.

INTRODUCTION

As solid oxide fuel cell becomes as the future electricity generation devices, its development is greatly explored. Currently, focus has been given to the materials, structures, and fabrications designs of SOFC components. Besides that, production cost and manufacturing processes were also considered so that the performance of SOFC can be enhanced to meet the requirement of economic

power production. Besides that, production cost and manufacturing processes were also considered so that the performance of SOFC can be enhances to meet the requirement of economic power production.

The most common structural designs of SOFCs are planar and tubular designs

[*] **Corresponding author Mohd Hafiz Dzarfan Othman:** Advanced Membrane Technology Research Centre (AMTEC), Universiti Teknologi Malaysia, 81310 Skudai, Johor, Malaysia; Tel.: +607 5536373, Fax: +607 5535925; E-mail: hafiz@petroleum.utm.my

Moisés R. Cesário & Daniel A. de Macedo (Eds.)

which both designs have different shapes and structures. Due to planar design caused thermal shock problem, the tubular design has been introduced by extrusion technique to overcome the issue. However, as inverse tubular cell diameter affects the power density, efforts had been made by Kendall [1] to develop micro-tubular SOFCs (MT-SOFCs) that possess smaller tubular diameter and enhanced performance. In fact, this developed advanced cell design offers a quick start-up capability, an excellent thermal stability during rapid heat cycling, low capital cost, high power output density, and easy to mobile as compared to the conventional tubular and planar SOFCs [2].

Typically, the design of SOFCs can be constructed into either self-supporting or supported concept, as depicted in Fig. (1) [3]. From Fig. (1), electrolyte with thickness of ca. 80-250 μm is denoted as self-supporting geometry which acts as structural element. For the supported design, a thin layer electrolyte with approximate thickness of <50 μm is deposited on cathode or anode which provide the mechanical strength to the hollow fiber. Depending on the support, the MT-SOFC materials can be constructed as anode, cathode or even electrolyte supported. Unlike electrolyte self-support, the use of anode-supported design possessed thinner electrolyte layer, and thus could offer a minimal electrolytic resistance losses and better production of conductance at lower temperatures [4]. Despite this design has excellent performance, precautions are needed to ensure the electrolyte layer is microcracks-free and completely gas-tight. Nonetheless, it should be noted that most of the recent works on the MT-SOFCs have developed the anode-supported design.

The recent trend in MT-SOFC development is to operate the system at lower temperature which is intermediate temperature (500-700 °C) or lower (< 500 °C). Some drawbacks of commercialized high operating temperature system are durability of the cell components, limitation of the material selections, as well as high production cost. On the other hands, lower temperature operation offers enhanced durability, quicker start-up, and greater robustness, wider choice of selecting materials with simplified system requirements.

Other than that, lowering the operating temperature from 700-1000 °C to 500-700 °C has become as one of the major tasks in the development of SOFCs with improved performance. Since last decade, electrolytes with high oxygen-ion conductivity operated at low temperatures have received tremendous attentions. One of the good option of alternative electrolyte to yttria stabilized zirconia (YSZ) is Cerium gadolinium oxide (CGO). Like zirconia, ceria-based electrolytes have a cubic, fluorite-type crystal structure. However, it has superior oxygen ion conductivity at low temperatures when compared to zirconia [6]. These features resulted to large oxygen ions conductivity with excellent mobility. Nevertheless, ceria will be partially reduced to be an electronic conductor when it is exposed to higher temperatures and at low oxygen partial pressures, which can cause short circuit [7]. Although the stability of ceria at low oxygen partial pressures is weaker than zirconia, its chemical stability of ceria with cathode materials is good. CGO has been proven with a wide variety of electrodes, including lanthanum strontium cobaltite (LSC), lanthanum nickel ferrite (LNF), lanthanum strontium manganite (LSM), lanthanum strontium ferrite (LSF) and lanthanum strontium cobalt ferrite (LSCF) [8].

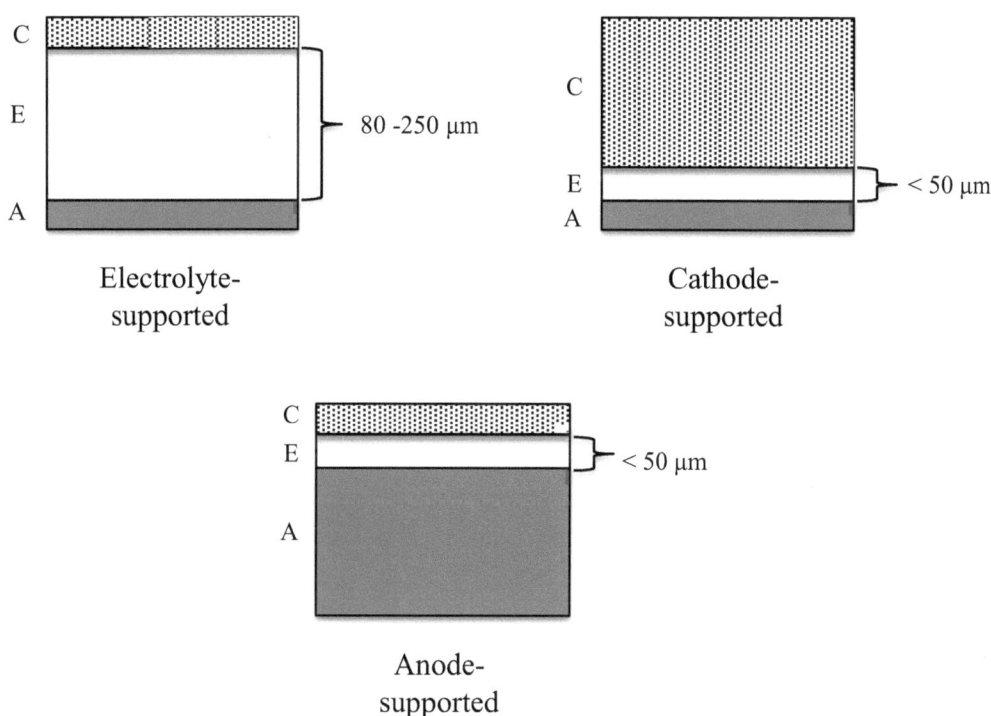

Fig. (1). Geometries of SOFCs (Reproduced from [5]).

However, the fabrication processes of MT-SOFCs, specifically the fabrication parameters that affect the structures of electrolytes fabricated *via* phase-inversion based technique is fairly reported. Micro-tubular that has smaller cell size led to a more challenging fabrication procedures as compared to the typical tubular design. Therefore, in the context of this chapter, discussion is focused on the fabrication of electrolytes *via* phase-inversion technique. The discussion has been divided into three sub-sections. Section one mainly describes on the concept of phase-inversion technique. Then, section two focuses on parameters that affect the fabrication of electrolyte-supported hollow fiber while the last section discussed on the role of a thin layer electrolyte in MT-SOFC application. Moreover, the fabrication parameters affecting the physical properties and electrochemical performances of electrolytes were discussed as well.

Phase-Inversion Based Extrusion/Sintering Technique for Fabrication of Ceramic Hollow Fiber

Phase-inversion based extrusion/sintering technique has been extensively implied for fabricating ceramic hollow fiber membranes to be used in numerous applications. There are three major steps involved using this technique: 1) preparation of spinning suspension, 2) extrusion of ceramic hollow fiber precursors, 3) sintering process [9]. Each of the steps plays a crucial part in the production of desired ceramic hollow fibers, as depicted in Table **1**.

Table 1. Step of phase-inversion based extrusion/sintering, their factors and influences to the hollow fiber properties.

Steps	Factors	Influences
Preparation of spinning suspension	• Concentration of compounds • Particle size/distribution Particle/binder ratio • Solvent types • Dispersant • Additives • Mixing speed (shear stress) • Mixing temperature	• Homogeneous • Viscosity • Particle packing density Rheology • Particle dispersion
Extrusion of ceramic hollow fiber precursors	• Suspension viscosity/homogeneous • Particle size/distribution • Extrusion speed • Air gap • Environment conditions (humidity/temperature) • Spinneret dimension/ configuration • Internal/external coagulant (solvent %, flow rate, temp.) • Suspension temperature	• Uniformity (bore shape) • Particle packing density • Surface morphology (smooth/rough) • Precursor morphology (dense/porous) • Precursor structure (symmetric/asymmetric) • Precursor thickness Precursor dimension Precursor length
Sintering process	• Particle characteristics (thermal expansion Particle size/distribution • Sintering profile Sintering placement (vertical/horizontal) • Environmental gas	• Membrane dimension Membrane morphology (dense/porous) • Cracking/defect • Grain size • Porosity • Pore size • Mechanical strength

In phase-inversion process, the ceramic material in the form of the hollow fiber precursor is resulted from the exchange between the solvent and the non-solvent (coagulant) which stimulates the precipitation of polymer in the suspension [10]. The basic concept of this process can be described by using the ternary phase diagram of polymeric system, encompassing polymer, solvent and non-solvent, as shown in Fig. (**2**). The entire phase-inversion process of a polymeric solution is represented by the path from A to D. Point A refers to the original polymeric solution, where no precipitation agent (non-solvent) is present in the solution. Solvent will diffuses out of the polymer solution, while non-solvent diffuses into the solution, during the immersion of polymeric solution into a non-solvent coagulation bath. If the flux of solvent is higher that the flux of non-solvent, the polymer concentration at interface would increase, simultaneously, the polymer starts to precipitate (denoted by point B). Point C, solidification of the polymer-rich phase is the result from continuous replacement of solvent by non-solvent. Any further solvent/non-solvent exchange caused to the shrinkage of polymer-rich phase that leads to point D where two phases (solid and liquid) are at equilibrium. The membrane structure of a solid (polymer rich) phase and the liquid (polymer poor) phase in which non-solvent will filling the membranes pores are denoted by point S and point L, accordingly.

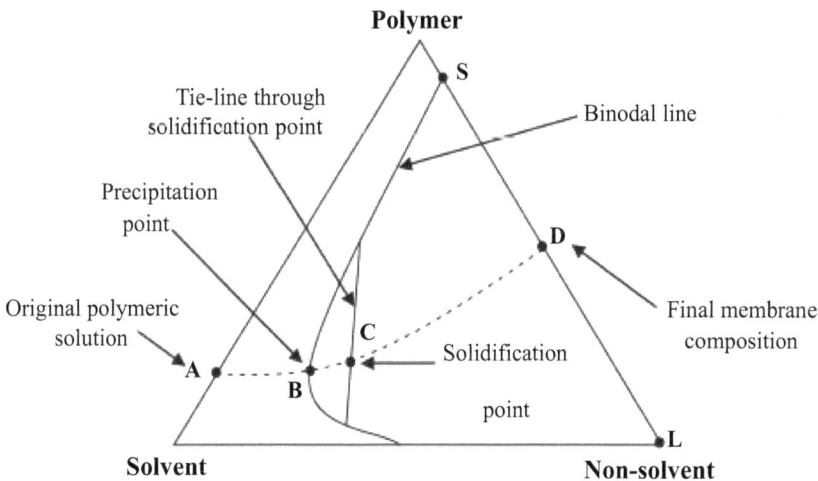

Fig. (2). Schematic ternary phase diagram of polymer/solvent/non-solvent during polymeric membrane formation (Reproduced from [11]).

In the phase-inversion process, the types of polymer binder, solvent or non-solvent indeed could affect the precipitation mechanism. For an instance, the N-Methyl-2-pyrrolidone (NMP) and dimethyl sulfoxide (DMSO) are solvents which their systems possess different precipitation line. As illustrated by Fig. (**3**), the precipitation point for the polyethersulfone (PESf)/DMSO/water system is closer to the original casting solution (0% water) line than that of the PESf/NMP/water system. Since the rates of solvent/non-solvent exchange between both systems are not significantly differ, the duration to reach the precipitation point for PESf/DMSO/water system is shorter than PESf/NMP/water system. Due to precipitation point largely determined the macrostructure of the membrane; PESf/DMSO/water and PESf/NMP/water systems would yield different macrostructures.

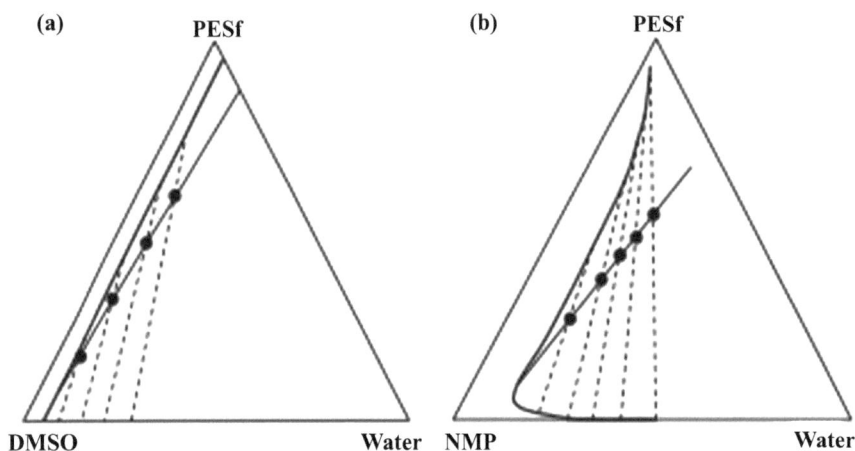

Fig. (3). Phase diagrams for (**a**) PESf/DMSO/Water and (**b**) PESf/NMP/Water systems. Thick line, binodal line; thin line, spinodal line; dashed lines, tie-lines; dot, solidification point (Reproduced from [12] and [13]).

Extensive studies have been conducted in order to understand and control the formation mechanisms for the wide-ranging structures observed in polymeric membrane formation [13, 14]. However, this knowledge is of inadequate use during ceramic membrane preparation ascribed by the large differences between polymeric and ceramic systems, specifically the low polymer concentration. As a matter of fact, only two morphologies have been observed in ceramic systems, namely finger-like voids and sponge-like structure, as shown in Fig. (**4**).

In ceramic membrane, the formation of finger-like voids precursors can be explained *via* a well-known phenomenon called hydrodynamically unstable viscous fingering. This phenomenon occurs at the interface between fluids with different viscosities at the initial time of mixing [16]. When spinning suspension is in contact with non-solvent, a steep concentration gradient will results to solvent/non-solvent exchange, a fast increase in term of local viscosity that eventually precipitates the polymer phase. It should be noted that instabilities at the interface between the suspension and the precipitant will cause the occurrence of viscous fingering that initiates the formation of finger-like. Under typical conditions, the two phases of different viscosities would establish a stable interface; however, the presence of the invertible polymer binder leads to a rapid increase of viscosity followed by polymer precipitation which preserved the viscous fingering structures. This relative thickness of finger-like structures and greatly influence the properties of membrane/membrane support in term of mechanical strength and permeation flux, attributed by the flexibility of ceramic hollow fiber. Thus, it is vital to tailor the morphology of fibers so that it can be used for any specific applications where phase-inversion technique is superior to be used to achieve the desired internal morphologies [9, 16].

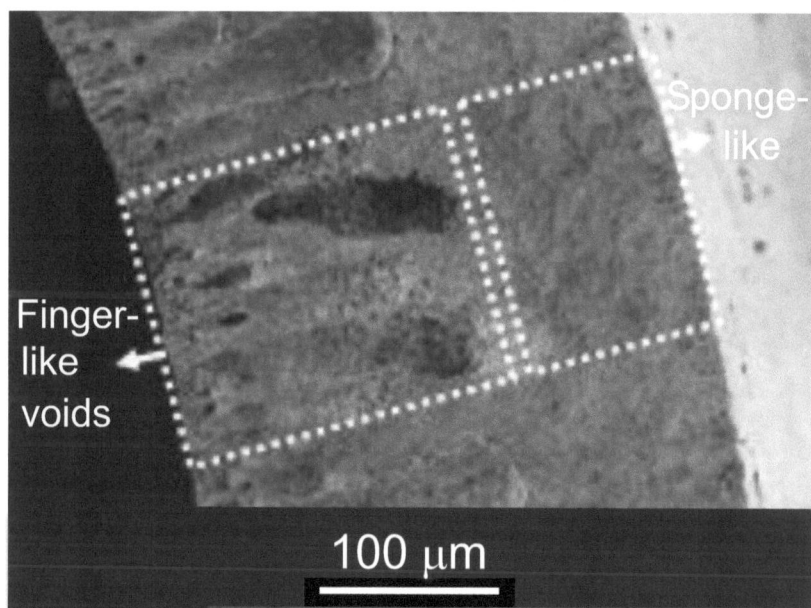

Fig. (4). Example of the macrostructure of ceramic hollow fiber precursor (Data from [15]).

Electrolyte-supported Hollow Fiber

Electrolyte is the heart of the fuel cell where it determines the performance of a cell. The structure of electrolyte must be dense so that it can acts as membrane that separates the air and fuel compartments and it must be a good oxygen ion conductor. Therefore, an electrolyte should possesses these following criteria to be as an excellent electrolyte: 1) high ionic conductivity, 2) low electronic conductivity, 3) stable in both oxidizing and reducing environments and 4) good mechanical characteristic [7, 8, 17, 18]. For SOFCs, many studies have reported about YSZ and CGO as a good candidate of electrolyte system [8, 18, 19]. Table

2 shows the general comparison between YSZ and CGO system along with their advantages and drawbacks.

As shown in Table **2**, YSZ can be operated at high temperatures (800-1000 °C) with sufficient mechanical strength. However, it is reactive towards its perovskite oxide electrodes such as LSM which this limits the wide use of YSZ [20]. At high temperatures, they react and forming pyrochlore, $La_2Zr_2O_7$, the perovskite $SrZrO_3$ or both [7] except the YSZ layer is shielded with CGO layer [18]. Alternatively, CGO has received great attention as electrolyte material since it offers lower polarization resistance and the highest conductivity at lower temperature [21]. Besides, doped ceria is more stable ad relatively unreactive toward other potential electrode materials, as compared to YSZ.

Table 2. Comparison between YSZ and CGO systems.

Electrolyte system	YSZ	CGO
Operating Temp	High Temp (800 – 1000 °C)	Intermediate Temp (500-700 °C)
Material Cost	Less expensive	Expensive
TEC	10.5×10^{-6} K^{-1}	12.5×10^{-6} K^{-1}
Conductivity	Ionic conductivity	Mixed ionic electronic conductivity
Mechanical strength	Sufficient mechanical strength	Ceria electrolytes are unstable mechanically at temperatures above about 700 °C
Reaction with electrodes material	Chemical reaction with the cathode material	Unreactive towards potential electrode materials
Advantages	High quality exhaust heat	Improved durability, wider choice of material selection, and lower costs

Based on literature findings, studies on electrolyte-supported SOFC are mainly focused on the development of YSZ hollow fiber. Li *et al.* fabricated YSZ hollow fiber membranes prepared by combining phase-inversion and sintering techniques by where the thickness of the membrane's dense layer is approximately around 120µm [22, 23]. One significant advantage of the phase-inversion process is the internal structure of the hollow fiber can be tailored. However, the study did not demonstrate its electrochemical performance. Few years later, same research group, Wei and Li [24] reported the fabrication of YSZ ceramic hollow fibers as an electrolyte-supported MT-SOFC. In their study, thin YSZ hollow fiber electrolyte-supported membranes with improve mechanical strength and gas-tight properties was fabricated where this study also firstly reported the electrochemical performance. The membranes possessed asymmetrical structures with porous inner surface and thin dense later near the outer wall, as shown in Fig. (**5**).

The inner surface of the anode was deposited with nickel using electroless plating, whereas the outer surface of the fiber was dip-coated with LSCF as the cathode layer were dip-coated prior of sintering process. Maximum power density produced by the cell was 0.018 Wcm^{-2} at 800 °C. Despite the thicker electrolyte (120 µm) improved the mechanical strength up to 297 MPa, the electrolyte-supported SOFCs still suffered from high electrolyte ohmic loss, which finally lowered the power density. Hence, reducing the thickness of the electrolyte is important so that ohmic loss can be avoided, and at same time sufficient mechanical strength can be maintained. Moreover, similar to LSM, the usage of LSCF as a cathode might affect the YSZ system but not to a significant extent, probably due to the lower cobalt oxide activity credited by the existent of ferrite in LSCF [7].

Fig. (5). Cross section SEM image of the YSZ electrolyte of hollow fiber after sintering (Data from [24]).

Apart from that, a highly asymmetric YSZ hollow fibers, which consisted of thin dense skin and porous layers was prepared by Yin *et al.* [25] by modifying the phase-inversion and sintering technique using a mixture of N-methyl-2-pyrrolidone (NMP) and water as internal coagulant. The thin dense skin layer serves as electrolyte film while the porous layer acts as an electrode of the SOFC by impregnating or electroless-plating of catalysts into its pores. Such microstructure can be obtained *via* a simple modification by mixing a solvent and water as the internal coagulant while untreated municipal water as the external coagulant. Fig. (6) shows the SEM micrographs of YSZ hollow fiber precursors under different magnifications. It can be seen that the finger-like pores were formed elongated across the membrane inner structure and thin skin layer of YSZ of 5.4-7.1 µm thickness. This highly asymmetric structure formation might be attributed to the different precipitation rates between the outside and inside the hollow fiber during the solvent/non-solvent exchange [26].

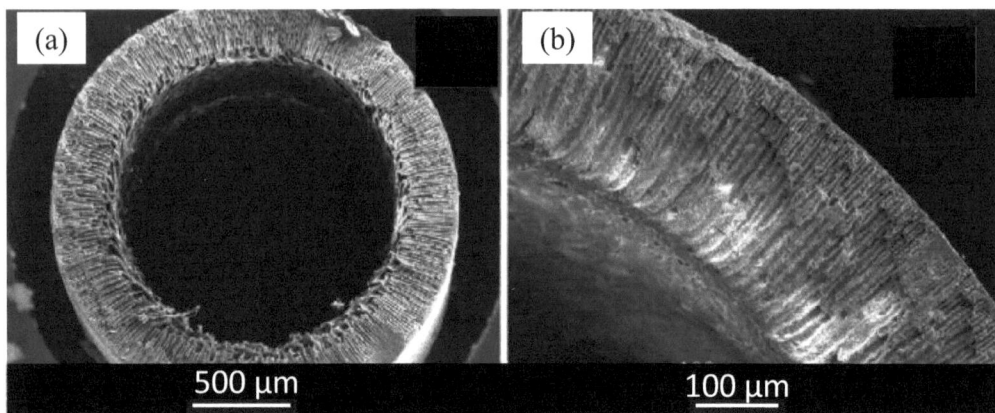

Fig. (6). SEM micrographs of the YSZ hollow fiber precursors: (**a**) cross-sectional; (**b**) membrane wall (Data from [25]).

The development of hollow fibers based YSZ electrolyte was further studied by Li and co-workers [27, 28]. At this point, focus of research is specified on the optimization of Ni electroless plating formula, increasing the deposition of Ni particles within the YSZ micro-tubes by utilizing a highly asymmetric electrolyte microstructure and characterizing the anode and cathode electrodes. The optimized electroless plating recipes were firstly tested on YSZ plates [27] and then on YSZ fibers [28]. As shown in Fig. (7), the microstructure of fibers was controlled by varying the fabrication parameters. Optimizing these parameters resulted to open and connected pores on the external and internal surfaces for a high specific surface area of the electrodes, and a thin dense gas-tight central layer for low ohmic losses. With the total wall thickness of 300 µm and external diameter of 1.6 mm, this fiber exhibited a poor performance with the maximum power density of only 0.0019 Wcm^{-2} at 800 °C. Nevertheless, it is being claimed that charge transfer or mass transport in the anode has become the major factor of these losses.

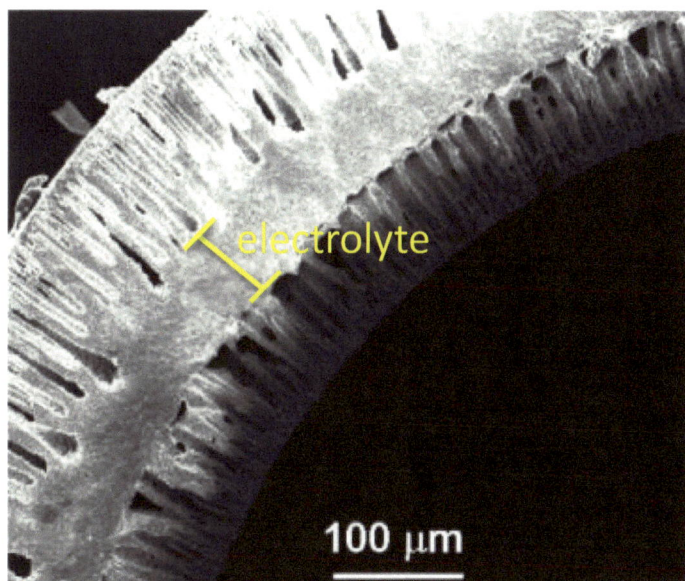

Fig. (7). Cross section SEM image after Ni electroless plating by dipping (Data from [28]).

Based on their previous study, the same research group investigated the effects of such microstructure to the performance of the cell by using highly asymmetric YSZ fiber with approximately 10 μm electrolyte thickness [29]. The fiber was prepared by infusing a mixture of NiO-YSZ particles into fiber pores using alcoholic dispersions and sintered at 1300 °C. The inner surface was deposited with Ni by electroless plating method to increase the anode conductivity while the LSM cathode was brush painted on the outer surface. However, several drawbacks related to the deposition step on the internal electrode were detected upon fabrication of these highly asymmetric YSZ hollow fibers. For example several cycles of deposition and sintering are required for the impregnation of the anode, which is time consuming and costly, especially when multiple cells were assembled. Several studies have proposed some solutions to overcome this limitation, yet the poor performance and conductivity of the electrode was observed due to lower thickness, higher porosity and limited control over the microstructure. As a result, the initial SOFC performance test with 5% H_2 as the fuel and air as the oxidant resulted in maximum power densities of only 0.018 Wcm^{-2} at 800 °C. This low value is mainly affected from the discontinuous nature of the electroless deposited nickel anode as shown in Fig. (**8**).

Studies on hollow fiber SOFCs using phase-inversion and sintering techniques paved the way for the other teams in China led by Tan and Yang to study on similar material. Their first attempt was fabricating electrolyte-supported cell.

This published study [30] reporting the use of phase-inversion an sintering technique for the fabrication of scandia doped ceria (SDC) tube with an outer diameter of 2 mm and 180 μm thick. Barium strontium cobalt ferrite (BSCF) acts as the cathode and the anode consisted of SDC-NiO where both electrodes were deposited inside and on the outer surface, respectively. Their study suggested that the sintering temperature should be higher than 1450 °C to obtain completely gas-tight micro tubes. Increasing sintering temperature increased the mechanical strength of the micro tubes up to 208 MPa when sintered at 1500 °C. The complete BSCF/SDC/NiO-SDC cells showed a stable power density of 0.106 Wcm^{-2} at 750 °C with pure H_2 as fuel and air as oxidant, accordingly.

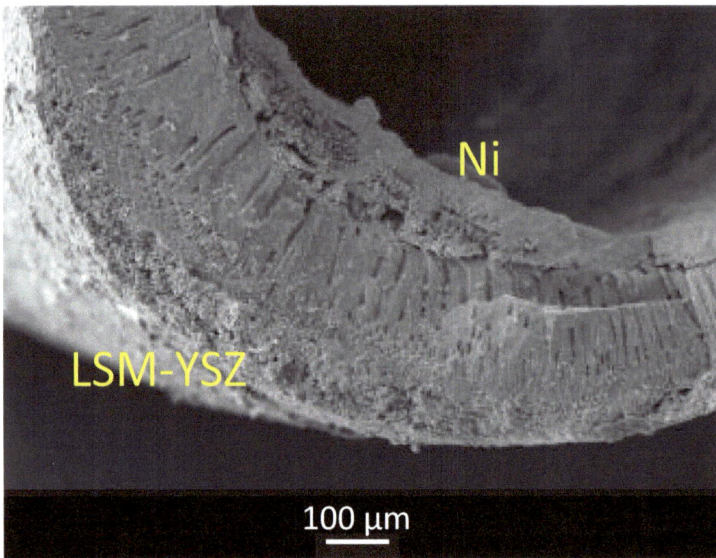

Fig. (8). SEM cross section image of partially assembled SOFC. (Data from [29]).

In addition, the thickness of the dense layer of the electrolyte should be optimized in order to enhance the output performance of the MT-SOFCs, as well to reduce the loss of ohmic while concurrently maintaining the required mechanical strength to support the electrodes. The porous layer might provide high surface area to accommodate one of the two electrodes. Electrolyte-supported MT-SOFC has longer-lasting durability, can survives high thermal, redox cycling tests, and higher mechanical strength as compared to other fuel cell design, however its thick electrolyte layer induces high ohmic losses and thus making the cell delivers poor power densities.

Electrolyte as Thin Layer

In order to enhance the performance of MT-SOFC while preserving maintaining mechanical strength, the thickness of electrolyte must be reduced. Thus, the anode-supported MT-SOFC could become as one of the preferable design to achieve the above features. This type of configuration is more efficient than the others due to some reasons. Firstly, its thickness is controllable (resulting in a thickness of 5-15 μm) since the electrolyte layer is deposited after extrusion of the thicker anode. Furthermore, since the sintering profiles of anodes and electrolytes can be matched, the electrolyte layer can be co-sintered with the anode substrates. This thin electrolytes increases the power densities while the co-sintering process reduces the entire manufacturing time and cost.

With the regard of electrolyte-supported hollow fiber MT-SOFCs that has been previously discussed, Li and co-workers started to develop anode-supported MT-SOFCs [31 - 33]. In the meantime, phase-inversion technique was modified to the co-extrusion/co-sintering technique to prepare dual-layer MT-SOFCs where two layers of anode and electrolyte were produced in a single-step, as shown in Fig. (**9**) [34].

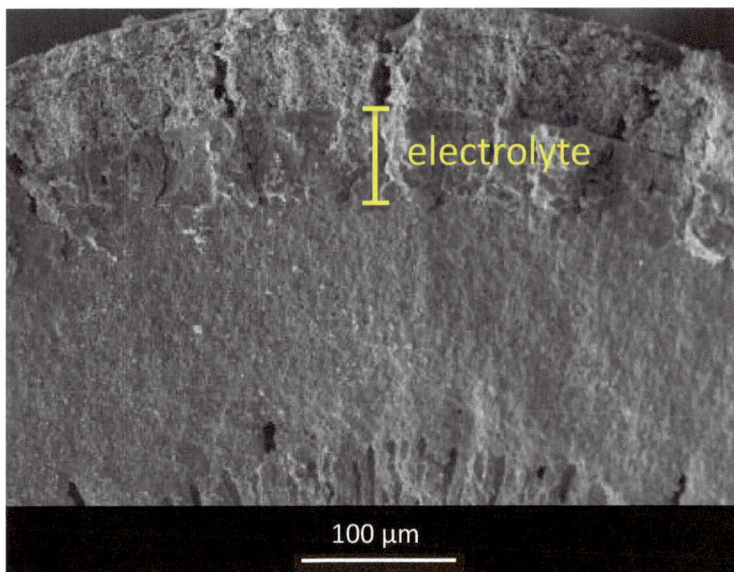

Fig. (9). Cross-sectional SEM image of the micro-tubular HF–SOFC showing anode/electrolyte/cathode layers after anode reduction (Data from [34]).

The co-extrusion of the CGO/Ni-CGO hollow fibers was accomplished by using a triple orifice spinneret. The systematic diagram of the co-extrusion process is shown in Fig. (**10**), while the photographic images and dimension of the spinneret are shown in Fig. (**11**). However, with electrolyte thickness of 80 μm the fuel cell only produced low power densities, which was approximately 0.08 W cm^{-2} at 550 °C.

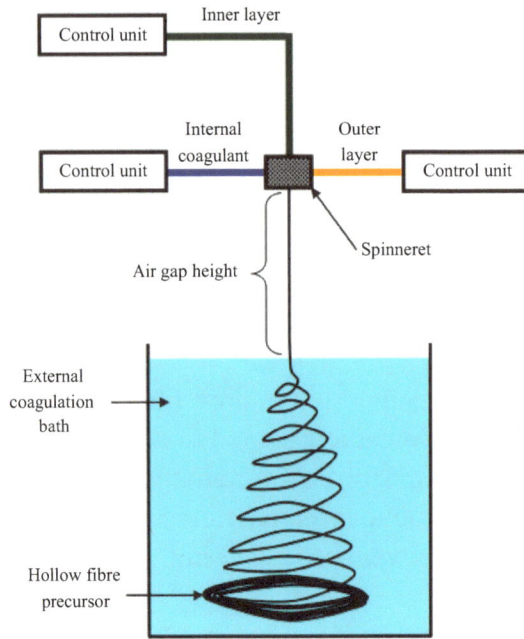

Fig. (10). Schematic diagram of the phase-inversion-based co-extrusion process of the dual-layer hollow fibers precursor (Data from [33]).

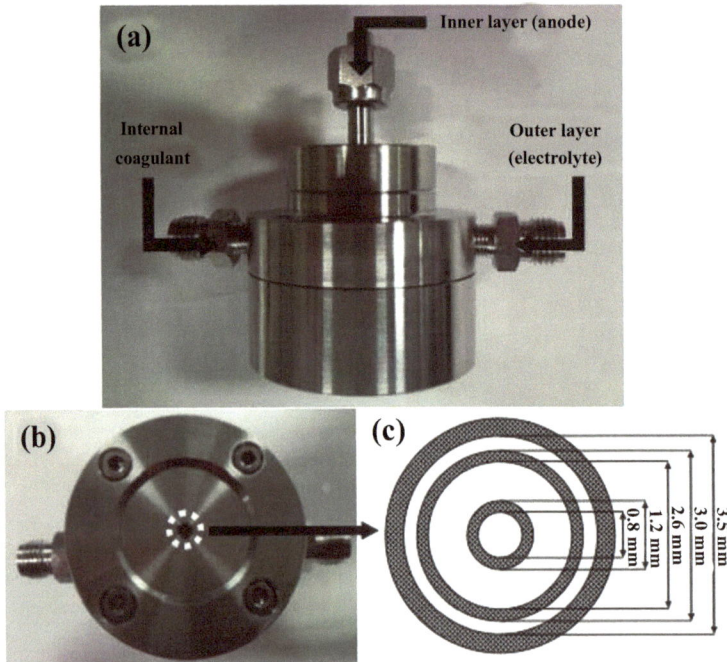

Fig. (11). Design and dimensions of the triple orifice spinneret (Data from [33]).

To increase the power output of the cell, Othman *et al.* [35] attempted to reduce the thickness of the electrolyte layer by reducing the extrusion rate. As shown in Fig. (**12**), reducing from 5.0 to 0.5 cm^3min^{-1} led to the hollow fiber SOFC with electrolyte layer thicknesses varying from 70 to 10 μm. A significant enhancement in term of power output of the cell, maximum power density up to 1.11 Wcm^{-2} was achieved by with the thinnest electrolyte layer (*i.e.* 10 μm), as presented in Fig. (**13**). The bending strength and the gas-tightness properties were also slightly reduced when the thickness of electrolyte was reduced. Although this configuration is generally enhanced the entire performance of MT-SOFCs, additional caution is still needed to prevent the formation of any microcracks as well as to ensure that the electrolyte is completely gas-tight.

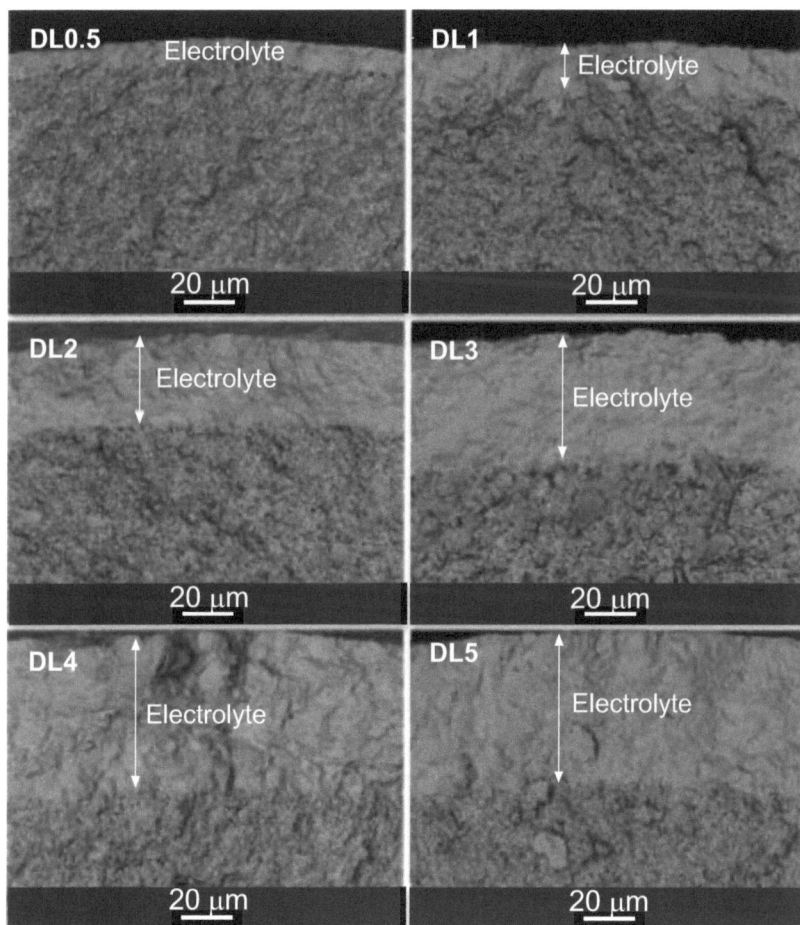

Fig. (12). Close-up SEM images of the electrolyte of the dual-layer HFs (Data from [35]).

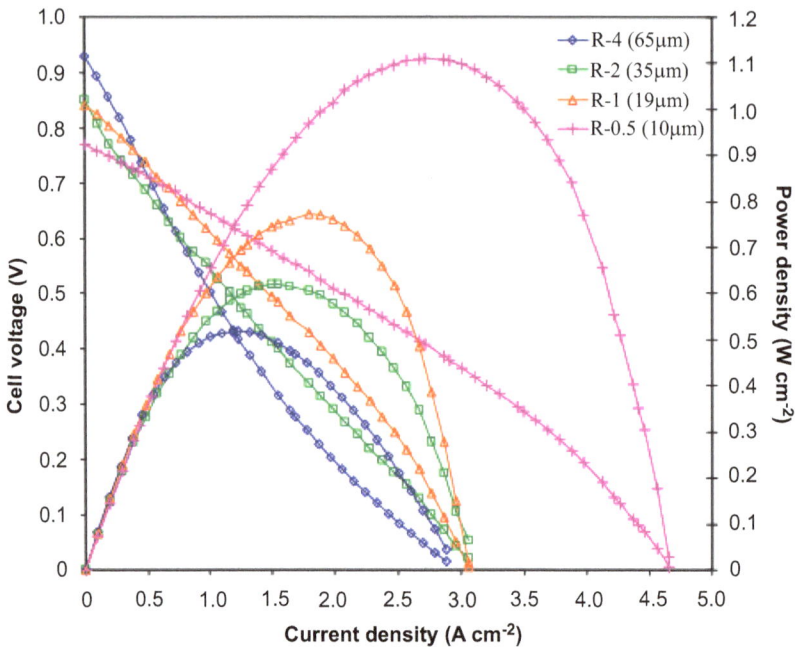

Fig. (13). Effect of current density on terminal potential differences and power densities for HF-SOFCs with different electrolyte thickness (Data from [35]).

Apart from that, Meng *et al.* [2] successfully fabricated an asymmetric YSZ/LSM–YSZ dual-layer cathode-supported hollow fibers *via* the co-extrusion/co-sintering technique. The dual-layer cathode supported hollow fibers consisted of dense YSZ on top layer supported by the LSM-YSZ substrate layer. Dual-layer cathode-supported hollow fibers consisting of a dense YSZ top layer supported on the LSM–YSZ substrate. Increasing the sintering temperature caused an improvement of the mechanical strength and compactness of the electrolyte later while the porosity of the cathode layer was decreased. In order to obtain suitable hollow fiber for electrolyte/cathode half cells, control of sintering temperature between 1350 and 1400 °C is required. From their study, the MT-SOFCs were fabricated by dip-coating the outer surface of the prepared dual-layer hollow fiber with 60 wt% NiO–YSZ, as shown in Fig. (**14**). However, the prepared electrolyte with thickness of 53 μm caused MT-SOFCs to exhibit a low power density of 0.29 W cm^{-2} at 850 °C, which high polarization resistance of the cathode could be the major reason that limits the performance of the fuel cells.

Besides that, the YSZ/NiO-YSZ dual-layer hollow fibers with thin YSZ top layer integrated on the porous NiO-YSZ (60:40 in weight) support have been developed as well [36]. Fig. (**15**) represents the microstructure of the dual-layer hollow fiber.

In their study, the mixture of NiO-YSZ anode spinning suspension was tailored by adding ethanol as non-solvent. It can be seen that the anode support and electrolyte layers possessed thickness of about 180 mm and 12 μm, respectively. To fabricate micro-tubular SOFCs, LSM cathode consisted of 20 wt%-YSZ was deposited on the electrolyte surface *via* dip-coating method. Experimental results indicated that dual-layer hollow fibers contained 15-20 wt% ethanol showed desired microstructure and much optimized properties which are the porosity of 38-35%, the conductivity of 3000 Scm^{-1} at room temperature, and the bending strength of 180 MPa. Better performance was depicted by the MT-SOFCs fabricated from such hollow fibers as compared to cathode-supported ones, with maximum power density of 0.485 Wcm^{-2} at 850 °C with 20 mLmin^{-1} of H$_2$ as fuel and 30 mLmin^{-1} air as oxidant, respectively.

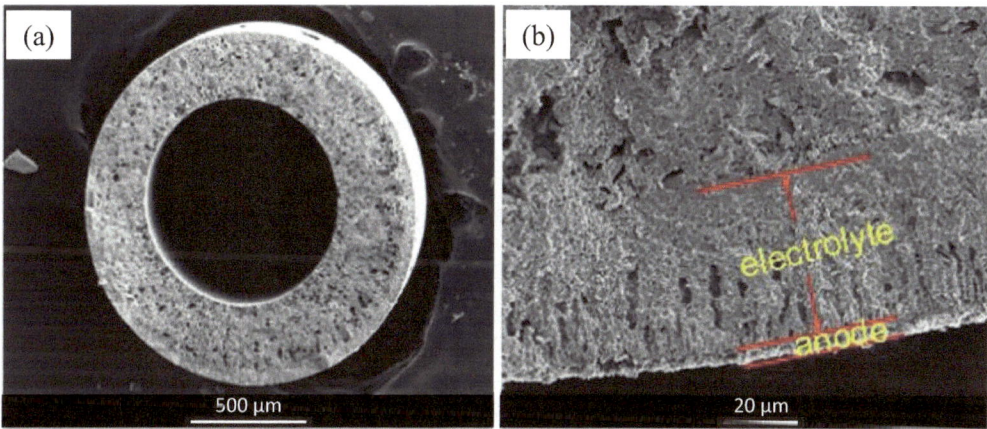

Fig. (14). Cross-sectional SEM images of the Ni-YSZ/YSZ/LSM-YSZ micro-tubular SOFC (**a**) cross-section; (**b**) magnified electrolyte and anode (Data from [2]).

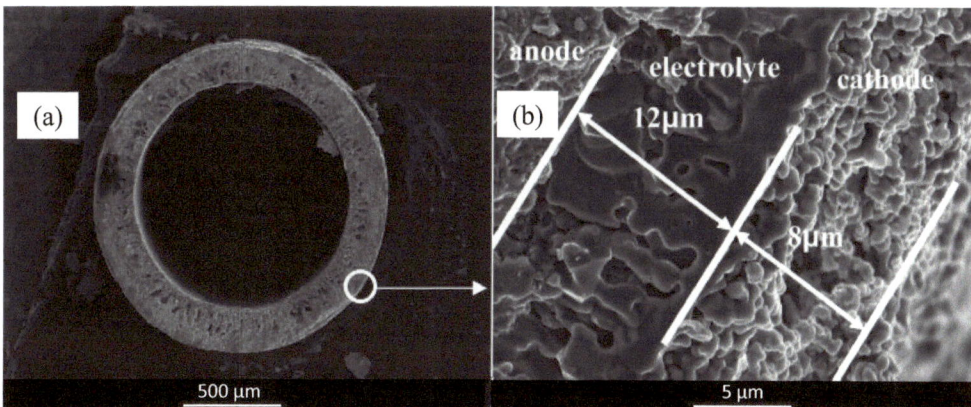

Fig. (15). Cross-section (**a**), and magnified electrolyte & cathode (**b**) of measured micro-tubular SOFC for 20 wt% in anode dope (Data from [36]).

Recently, Li *et al.* [37] conducted an advanced research on co-extrusion/c--sintering to fabricate triple-layer ceramic hollow fibers in a single step process using four orifice spinneret, illustrated in Fig. (**16**). The first two layers encompassed of 60%NIO-40%CGO anode and 40%NiO-60%CGO anode functional layer (AFL), meanwhile, the outer layer primarily is pure CGO electrolyte. Adjusting the extrusion rate of AFL during co-extrusion process resulted to AFL with different thickness and yet is greatly adhered between layers. As can be seen in Fig. (**17**), the precursor of the anode layers showed typical asymmetrical morphologies which are finger-like voids and sponge-like structures, these structures were perfectly preserved after the co-sintering process. Introduction of AFL between anode and electrolyte improved the mechanical strength and gas-tightness of the electrolyte. In addition, appropriate electrical conductivity has also being maintained, in which this suggests that continuous Ni path was developed within the hollow fibers. Nevertheless, finger-like voids were also observed to occupy the outer layer, this is unfavourable to the gas-tightness of electrolyte. Several modifications have been suggested to eliminate the voids such as increasing the flow rate of internal coagulant, increasing the viscosity of spinning and using a higher air gap.

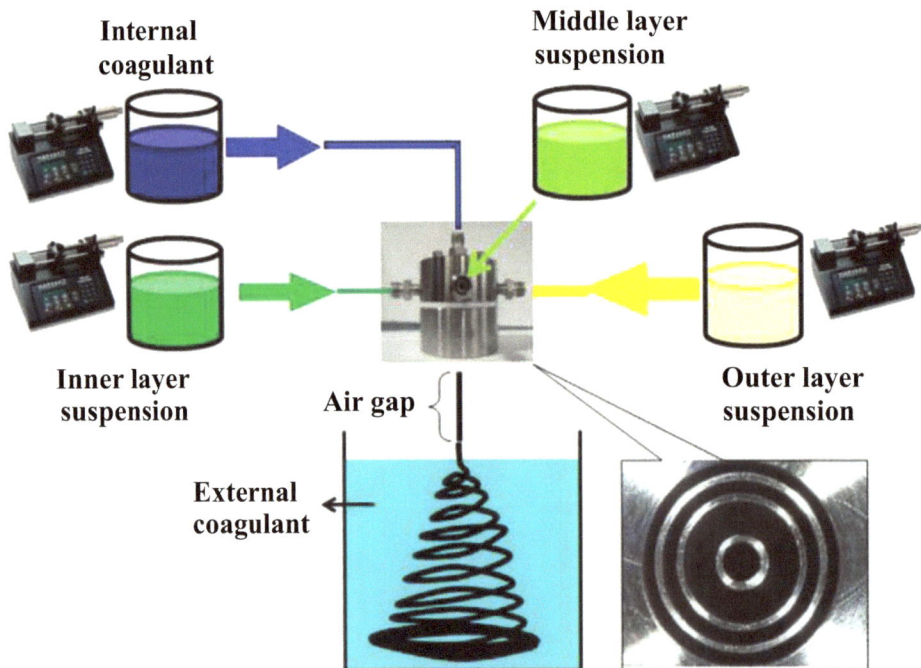

Fig. (16). Schematic diagram of the phase-inversion based co-extrusion process. The inset shows the picture of the quadruple-orifice spinneret (Data from [37]).

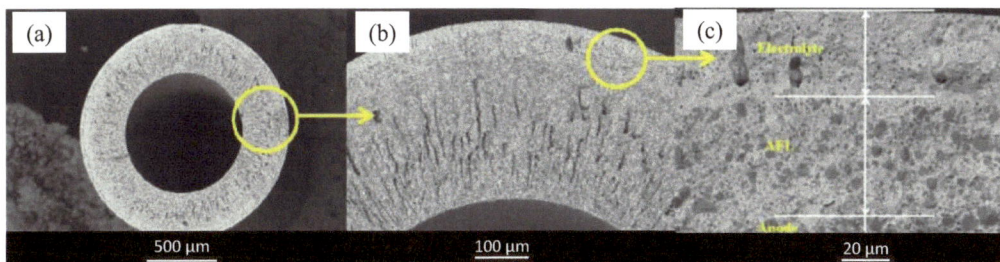

Fig. (17). SEM images (backscattered electrons (BSE) mode) of (**a**) whole view, (**b**) cross-section and (**c**) a higher magnification of cross-section of the sintered triple-layer hollow fibres, using the AFL extrusion rate at 3 ml/min (Data from [37]).

After several modifications has been made, they managed to conduct the electrochemical performance test on the modified triple-layer hollow fiber [38]. Study on developing more efficient cells was carried out by investigating the effects of AFL thickness on different physical and electrochemical properties in which the extrusion rates were accurately adjusted. As seen from SEM images in Fig. (**18**), the fibers displayed a typical asymmetrical structures consisting of sponge-like structures and finger-like channels when the triple-layer precursors was fabricated using quadruple orifice spinneret shown in Fig. (**19**). The construction of AFL has been found to effectively improve the gas-tightness of the electrolyte layer and the cell robustness. The most improved cell performance was obtained when AFL between 17 and 33 μm (6-11.5% of total anode thickness) with electrolyte thickness about 15 μm were used for constructing the cell. The maximum power density of the cell was 1.21 Wcm^{-2} of at 600 °C using pure H_2. This improvement is contributed by the balance existence of enlarging Triple phase boundary (TPB) and more mass transport resistance wen AFL with increased thickness was introduced.

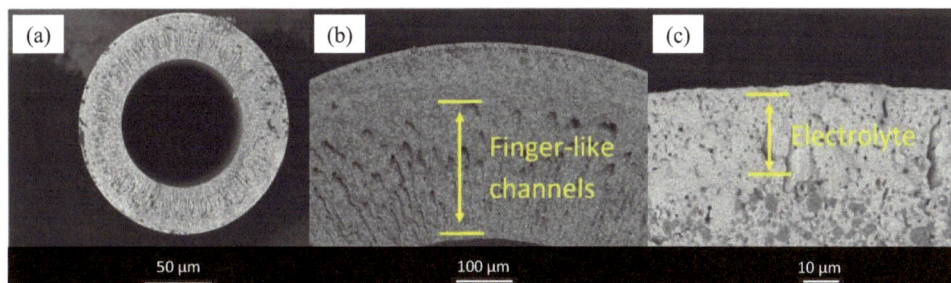

Fig. (18). SEM images of the fibre with the AFL extrusion rate of 2 ml min^{-1}: (**a**) overall view, (**b**) cross-section, (**c**) electrolyte/electrode interface (Data from [38]).

From the interpretations, the fabrication technique greatly influenced the electrolyte thickness. Phase-inversion based co-extrusion/co-sintering technique will typically induced asymmetrical structures consisting of finger-like voids and sponge-like region of fabricated anode supported MT-SOFCs. Both structures play their own role where finger-like structures provides a lesser resistance route for fuel gas and products transportation whereas sponge-like structures provides great number of TPB for electrochemical reactions and mechanical strength of the hollow fiber. In addition, this fabricating technique permits adjustments of extrusion parameters that allow the fabrication of controlled anode structures with desired thickness of electrolyte. Although high performance MT-SOFCs is also influenced by thickness of electrolyte, the structure of anode should be taken into account as well. Anode hollow fiber with longer finger-like voids thickness and high porosity structure affects the overall cell performance despite of higher gas permeation, the support's mechanical strength is strongly reduced and the TPB areas are ultimately destroyed. Thus, both anode and electrolyte need to be well structured in order to produce a high performance anode supported MT-SOFCs.

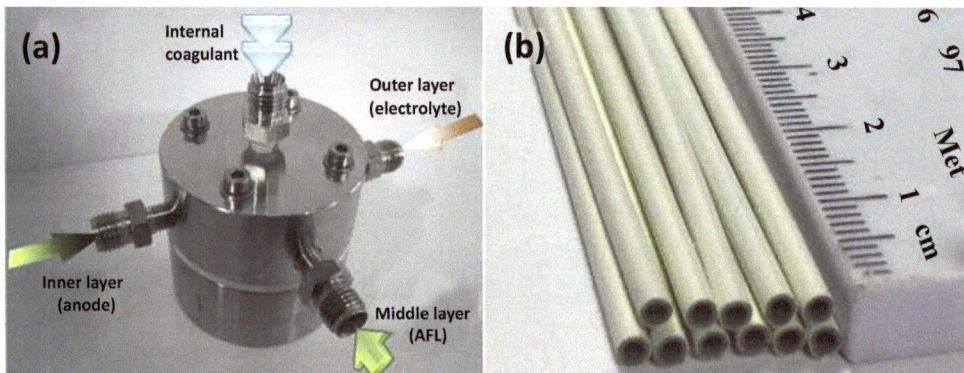

Fig. (19). Photographic pictures of (**a**) quadruple-orifice spinneret; (**b**) example of triple-layer precursors (Data from [38]).

CONCLUDING REMARKS

Implementation of phase-inversion based extrusion/sintering technique enable the control of internal structure of hollow fiber. The fabricated electrolyte hollow fibers typically have asymmetrical structures with porous inner surface and thin dense layer with thickness about 120 μm near the outer wall. However, high ohmic losses caused by the thick electrolyte layer will ultimately reduce the power densities. In order to reduce the ohmic losses while preserving the mechanical strength of electrodes support and to improve the output performance of the MT-SOFCs, the thickness of the dense layer of the electrolyte should be optimized. Recently, designation of the multi-layer tubing for micro SOFC applications particularly to produce thin layer electrolyte using anode-supported configuration can be achieved by phase-inversion based co-extrusion/co-sintering. As compared to conventional multi-step processes that require repeated coating and sintering process, this technique has demonstrated a number of desired characteristics such as better morphologies tailoring, great process control as well as reducing the fabrication costs. In addition, when using this technique, each component has better adhesion towards each other, prominent to reduce ohmic loss and contact resistance. Therefore, phase-inversion based co-extrusion/co-sintering has the potential to become an economical and reliable technique for mass-production.

CONFLICT OF INTEREST

The authors confirm that they have no conflict of interest to declare for this publication.

ACKNOWLEDGEMENTS

The authors gratefully acknowledge financial support from the Ministry of Education Malaysia under the Fundamental Research Grant Scheme (Project Number: R.J130000.7809.4F282), and Universiti Teknologi Malaysia under the Research University Grant Tier 1 (Project number: Q.J130000.2509.05H53). The authors would also like to thank Research Management Centre, Universiti Teknologi Malaysia for the technical support.

REFERENCES

[1]   Kendall, K.; Minh, N.Q.; Singhal, S.C. *High Temperature and Solid Oxide Fuel Cells*; Elsevier, **2003**.

[2]   Meng, X.; Gong, X.; Yang, N.; Tan, X.; Yin, Y.; Ma, Z-F. Fabrication of Y2O3-stabilize--ZrO2(YSZ)/La0.8Sr0.2MnO3−α–YSZ dual-layer hollow fibers for the cathode-supported micro-tubular solid oxide fuel cells by a co-spinning/co-sintering technique. *J. Power Sources,* **2013**, *237,* 277-284.
[http://dx.doi.org/10.1016/j.jpowsour.2013.03.026]

[3]   Badwal, S.P.; Foger, K. *Solid Oxide Electrolyte Fuel Cell Review,* **1996**, *8842*(95), 257-265.

[4]   Du, Y.; Sammes, N.M. Fabrication and properties of anode-supported tubular solid oxide fuel cells. *J. Power Sources,* **2004**, *136*(1), 66-71.
[http://dx.doi.org/10.1016/j.jpowsour.2004.05.028]

[5]   Othman, M.H. *High performance micro-tubular solid oxide fuel cell*; Imperial College, **2011**.

[6]   Lawlor, V.; Griesser, S.; Buchinger, G.; Olabi, G.; Cordiner, S.; Meissner, D. Review of the micro-tubular solid oxide fuel cell. *J. Power Sources,* **2009**, *193*(2), 387-399.
[http://dx.doi.org/10.1016/j.jpowsour.2009.02.085]

[7]   Fergus, J.W. Electrolytes for solid oxide fuel cells. *J. Power Sources,* **2006**, *162*(1), 30-40.
[http://dx.doi.org/10.1016/j.jpowsour.2006.06.062]

[8] Wincewicz, K.C.; Cooper, J.S. Taxonomies of SOFC material and manufacturing alternatives. *J. Power Sources,* **2005**, *140*(2), 280-296.
[http://dx.doi.org/10.1016/j.jpowsour.2004.08.032]

[9] Li, K. *Ceramic Membranes for Separation and Reaction*; John Wiley & Sons Ltd: England, **2007**.
[http://dx.doi.org/10.1002/9780470319475]

[10] Baker, R.W. *Membrane Technology and Applications*; John Wiley & Sons, Ltd: Chichester, UK, **2012**.
[http://dx.doi.org/10.1002/9781118359686]

[11] Strathmann, H.; Kock, K. The formation mechanism of phase-inversion membranes. *Desalination,* **1977**, *21*(3), 241-255.
[http://dx.doi.org/10.1016/S0011-9164(00)88244-2]

[12] B. F. Barton and a J. McHugh, "Kinetics of thermally induced phase separation in ternary polymer solutions. I. Modeling of phase separation dynamics. *J. Polym. Sci., B, Polym. Phys.,* **1999**, *37*(13), 1449-1460.
[http://dx.doi.org/10.1002/(SICI)1099-0488(19990701)37:13<1449::AID-POLB11>3.0.CO;2-T]

[13] Li, Z.; Jiang, C. Investigation of the dynamics of poly(ether sulfone) membrane formation by immersion precipitation. *J. Polym. Sci., B, Polym. Phys.,* **2005**, *43*(5), 498-510.
[http://dx.doi.org/10.1002/polb.20353]

[14] Barzin, J.; Sadatnia, B. Theoretical phase diagram calculation and membrane morphology evaluation for water/solvent/polyethersulfone systems. *Polymer (Guildf.),* **2007**, *48*(6), 1620-1631.
[http://dx.doi.org/10.1016/j.polymer.2007.01.049]

[15] Liu, S.; Li, K. Preparation TiO2/Al2O3 composite hollow fibre membranes. *J. Membr. Sci.,* **2003**, *218*(1–2), 269-277.
[http://dx.doi.org/10.1016/S0376-7388(03)00184-4]

[16] Kingsbury, B.F.; Li, K. A morphological study of ceramic hollow fibre membranes. *J. Membr. Sci.,* **2009**, *328*(1–2), 134-140.
[http://dx.doi.org/10.1016/j.memsci.2008.11.050]

[17] Veranitisagul, C.; Kaewvilai, A.; Wattanathana, W.; Koonsaeng, N.; Traversa, E.; Laobuthee, A. Electrolyte materials for solid oxide fuel cells derived from metal complexes: Gadolinia-doped ceria. *Ceram. Int.,* **2012**, *38*(3), 2403-2409.
[http://dx.doi.org/10.1016/j.ceramint.2011.11.006]

[18] Antoni, L. Materials for Solid Oxide Fuel Cells: the Challenge of their Stability. *Mater. Sci. Forum,* **2004**, *461–464*, 1073-1090.
[http://dx.doi.org/10.4028/www.scientific.net/MSF.461-464.1073]

[19] Lawlor, V. Review of the micro-tubular solid oxide fuel cell (Part II: Cell design issues and research activities). *J. Power Sources,* **2013**, *240*, 421-441.
[http://dx.doi.org/10.1016/j.jpowsour.2013.03.191]

[20] Chen, Y.; Yang, L.; Ren, F.; An, K. Visualizing the structural evolution of LSM/xYSZ composite cathodes for SOFC by in-situ neutron diffraction. *Sci. Rep.,* **2014**, *4*, 5179.
[PMID: 24899139]

[21] Othman, M.H.; Droushiotis, N.; Wu, Z.; Kelsall, G.; Li, K. High-performance, anode-supported, microtubular SOFC prepared from single-step-fabricated, dual-layer hollow fibers. *Adv. Mater.,* **2011**, *23*(21), 2480-2483.
[http://dx.doi.org/10.1002/adma.201100194] [PMID: 21484892]

[22] Liu, Y.; Chen, O.Y.; Wei, C.C.; Li, K. Preparation of yttria-stabilised zirconia (YSZ) hollow fibre membranes. *Desalination,* **2006**, *199*(1–3), 360-362.
[http://dx.doi.org/10.1016/j.desal.2006.03.216]

[23]   Liu, L.; Tan, X.; Liu, S. Yttria stabilized zirconia hollow fiber membranes. *J. Am. Ceram. Soc.,* **2006**, *89*(3), 1156-1159.
[http://dx.doi.org/10.1111/j.1551-2916.2005.00901.x]

[24]   Wei, C.C.; Li, K. Yttria-Stabilized Zirconia (YSZ) -Based Hollow Fiber Solid Oxide Fuel Cells. *Ind. Eng. Chem. Res.,* **2008**, *47*, 1506-1512.
[http://dx.doi.org/10.1021/ie070960v]

[25]   Yin, W.; Meng, B.; Meng, X.; Tan, X. Highly asymmetric yttria stabilized zirconia hollow fibre membranes. *J. Alloys Compd.,* **2009**, *476*(1–2), 566-570.
[http://dx.doi.org/10.1016/j.jallcom.2008.09.079]

[26]   Liu, S.; Tan, X.; Li, K.; Hughes, R. Preparation and characterisation of SrCe0.95Yb0.05O2.975 hollow fibre membranes. *J. Membr. Sci.,* **2001**, *193*, 249-260.
[http://dx.doi.org/10.1016/S0376-7388(01)00518-X]

[27]   Dal Grande, F.; Thursfield, A.; Metcalfe, I. Morphological control of electroless plated Ni anodes: Influence on fuel cell performance. *Solid State Ion.,* **2008**, *179*(35–36), 2042-2046.
[http://dx.doi.org/10.1016/j.ssi.2008.06.022]

[28]   Grande, F.D.; Thursfield, A.; Kanawka, K.; Droushiotis, N.; Doraswami, U.; Li, K.; Kelsall, G.; Metcalfe, I.S. Microstructure and performance of novel Ni anode for hollow fibre solid oxide fuel cells. *Solid State Ion.,* **2009**, *180*(11–13), 800-804.
[http://dx.doi.org/10.1016/j.ssi.2008.12.038]

[29]   Kanawka, K.; Grande, F.D.; Wu, Z.; Thursfield, A.; Ivey, D.; Metcalfe, I.; Kelsall, G.; Li, K. Microstructure and Performance Investigation of a Solid Oxide Fuel Cells Based on Highly Asymmetric YSZ Microtubular Electrolytes. *Ind. Eng. Chem. Res.,* **2010**, *49*(13), 6062-6068.
[http://dx.doi.org/10.1021/ie1002558]

[30]   Yang, N.; Tan, X.; Ma, Z.; Thursfield, A. Fabrication and Characterization of Ce 0.8 Sm 0.2 O 1.9 Microtubular Dual-Structured Electrolyte Membranes for Application in Solid Oxide Fuel Cell Technology. *J. Am. Ceram. Soc.,* **2009**, *92*(11), 2544-2550.
[http://dx.doi.org/10.1111/j.1551-2916.2009.03267.x]

[31]   Yang, N.; Tan, X.; Ma, Z. A phase-inversion/sintering process to fabricate nickel/yttria-stabilized zirconia hollow fibers as the anode support for micro-tubular solid oxide fuel cells. *J. Power Sources,* **2008**, *183*(1), 14-19.
[http://dx.doi.org/10.1016/j.jpowsour.2008.05.006]

[32]   Droushiotis, N.; Doraswami, U.; Kanawka, K.; Kelsall, G.H.; Li, K. Characterization of NiO–yttria stabilised zirconia (YSZ) hollow fibres for use as SOFC anodes. *Solid State Ion.,* **2009**, *180*(17–19), 1091-1099.
[http://dx.doi.org/10.1016/j.ssi.2009.04.004]

[33]   Othman, M.H.; Wu, Z.; Droushiotis, N.; Kelsall, G.; Li, K. Morphological studies of macrostructure of Ni–CGO anode hollow fibres for intermediate temperature solid oxide fuel cells. *J. Membr. Sci.,* **2010**, *360*(1–2), 410-417.
[http://dx.doi.org/10.1016/j.memsci.2010.05.040]

[34]   Droushiotis, N.; Othman, M.H.; Doraswami, U.; Wu, Z.; Kelsall, G.; Li, K. Novel co-extruded electrolyte–anode hollow fibres for solid oxide fuel cells. *Electrochem. Commun.,* **2009**, *11*(9), 1799-1802.
[http://dx.doi.org/10.1016/j.elecom.2009.07.022]

[35]   Othman, M.H.; Droushiotis, N.; Wu, Z.; Kanawka, K.; Kelsall, G.; Li, K. Electrolyte thickness control and its effect on electrolyte/anode dual-layer hollow fibres for micro-tubular solid oxide fuel cells. *J. Membr. Sci.,* **2010**, *365*(1–2), 382-388.
[http://dx.doi.org/10.1016/j.memsci.2010.09.036]

[36] Meng, X.; Gong, X.; Yin, Y.; Yang, N-T.; Tan, X.; Ma, Z-F. Microstructure tailoring of YSZ/Ni-YSZ dual-layer hollow fibers for micro-tubular solid oxide fuel cell application. *Int. J. Hydrogen Energy,* **2013**, *38*(16), 6780-6788.
[http://dx.doi.org/10.1016/j.ijhydene.2013.03.088]

[37] Li, T.; Wu, Z.; Li, K. Single-step fabrication and characterisations of triple-layer ceramic hollow fibres for micro-tubular solid oxide fuel cells (SOFCs). *J. Membr. Sci.,* **2014**, *449*, 1-8.
[http://dx.doi.org/10.1016/j.memsci.2013.08.009]

[38] Li, T.; Wu, Z.; Li, K. Co-extrusion of electrolyte/anode functional layer/anode triple-layer ceramic hollow fibres for micro-tubular solid oxide fuel cells-electrochemical performance study. *J. Power Sources,* **2015**, *273*, 999-1005.

Frontiers in Ceramic Science, 2017, *Vol. 1*, 131-163

Proton Conducting Ceramic Materials for Intermediate Temperature Solid Oxide Fuel Cells

Narendar Nasani, Francisco Loureiro and **Duncan Paul Fagg**[*]

Nanoengineering Research Group, Centre for Mechanical Technology and Automation, Department of Mechanical Engineering, University of Aveiro, 3810-193, Aveiro, Portugal

Abstract: Solid oxide fuel cells (SOFCs) offer a large potential as a future green energy technology, as they show high fuel conversion efficiency with limited pollution and fuel flexibility for a wide range of applications [1]. These devices operate in the high temperature range (800-1000 °C), where cost and longevity related problems have slowed their commercialization. Nonetheless, the high operating temperature hurdle of SOFCs can be crossed by using proton conducting ceramic oxides as solid electrolytes, producing so called protonic ceramic fuel cells (PCFCs). The perovskite $AB_{1-x}M_xO_{3-\delta}$, A=Ca, Ba, Sr; B=Ce, Zr; M=Y, Gd, Yb) materials have been suggested as electrolyte materials for PCFCs, since these materials show high protonic conductivity and lower activation energy (0.4-0.6 eV) in the intermediate temperature range 500-750 °C [2]. The alkaline earth doped cerate family possess high proton conductivity (about 10^{-2} S cm^{-1} at 600 °C) with low activation energy, but suffer from poor chemical stability due to degradation in the presence of acidic gases (*e.g.*, CO_2) and steam, precluding their practical use as electrolytes in PCFCs. On the contrary, the alkaline earth doped zirconates exhibit good chemical stability, although their overall proton conductivity is limited (about 10^{-3} S·cm^{-1} at 600 °C) due to a low grain boundary conduction, compounded by poor sinterability and limited grain growth [3]. By tailoring the chemical composition of these two material families, a compromise between good proton conductivity, good sinterability and chemical stability can be achieved [4]. Due to the novelty and potential of this technology, this chapter will be dedicated to recent developments in PCFCs, highlighting potential electrolyte and electrode materials, their microstructure and property relationships.

Keywords: Barium cerate, Barium zirconate, Cathode, Cermet anode, Combustion synthesis, Electrolyte, Perovskites, Proton conductors, Protonic ceramic fuel cell, Solid oxide fuel cell.

---

[*] **Corresponding author Duncan Paul Fagg:** University of Aveiro, Aveiro, Portugal; Tel: +351-234-370830; Fax: +351-234-370953; E-mail: duncan@ua.pt

Moisés R. Cesário & Daniel A. de Macedo (Eds.)

INTRODUCTION

Solid oxide fuel cells (SOFCs) are amongst the most capable technologies to deliver sustainable and clean energy to the future. These high-temperature targeted devices operate around 800-1000 °C, with high efficiency, zero emissions and silent function. The use of non-noble catalysts in the electrode materials and the possibility of co-generation are some of the advantages [5, 6]. However, several key technical issues related to the high operating temperature have hindered the deployment of this transformative technology, such as high systems costs and high performance degradation rates, as well as slow start-up and shutdown cycles [7]. The elevated operation temperature leads to severe degradation of cell components, and demand expensive sealants and interconnection materials [5, 6]. Over the past decade, considerable effort has, therefore, been focused in reducing the operating temperature of SOFCs down to the intermediate temperature range. Recently, innovative proton conductive electrolytes have opened a new working window, in the temperature range of 400-700 °C. These electrolytes derive their ionic conductivity from the incorporation of protonic defects of sufficiently high mobility [8].

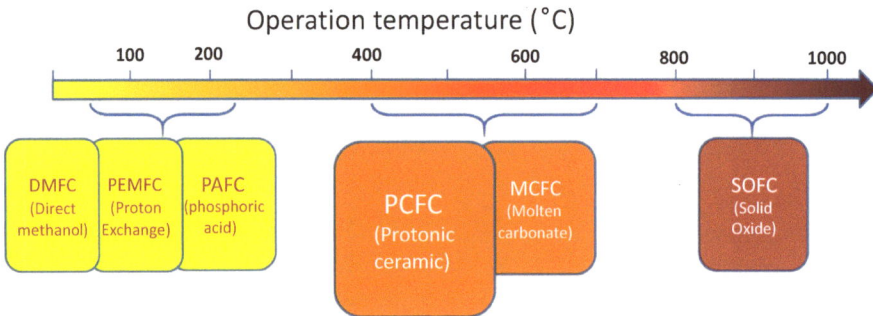

Fig. (1). Status of the current fuel cell technologies and their targeted operation temperature range.

Fuel cells can use a diverse set of electrolytes and it is this component that normally defines the operation temperature (Fig. **1**), the nature of the fuel and the kind of reforming demanded for the fuel (internal or external) [9]. In the low temperature range, the most advanced technology is by a wide margin the proton exchange membrane fuel cell (PEMFC), leaving a fringe role for the phosphoric acid fuel cell (PAFC) and the direct methanol fuel cell (DMFC). On the other hand, in the medium to high temperature range, the molten carbonate fuel cell (MCFC) and SOFC are mainly used for providing high power for stationary applications and often integrate cogeneration systems (production of electricity and heat together) [10]. Amongst these, the more recent PCFC technology is very promising because it offers the same flexibility of SOFC in terms of the type of fuels that can be used (*e.g.* hydrogen or hydrocarbons) and also does not contain corrosive liquids, as opposed to MCFCs for instance, while being able to be operated at intermediate temperatures [11].

A PCFC is comprised of a proton conducting solid electrolyte, sandwiched between two porous electrodes (the anode and the cathode) [12]. The working principle is illustrated in Fig. (**2**). In PCFCs, the fuel (hydrogen or alternative hydrocarbons) is oxidised at anode side by creating protons and releasing electrons:

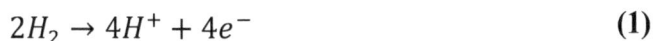

$$2H_2 \rightarrow 4H^+ + 4e^- \tag{1}$$

The protons formed at anode side migrate through the electrolyte towards cathode side where the formation of water takes place by reaction with oxygen:

$$4H^+ + 4e^- + O_2 \rightarrow 2H_2O \tag{2}$$

Conversely, in the case of oxide-ion conducting electrolytes, these mobile ions are formed at the cathode by oxygen reduction:

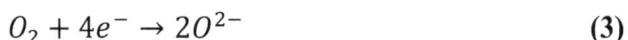

$$O_2 + 4e^- \rightarrow 2O^{2-} \tag{3}$$

While on the other side, the anodic reaction will be:

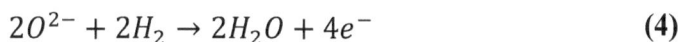

$$2O^{2-} + 2H_2 \rightarrow 2H_2O + 4e^- \tag{4}$$

Global reactions, equations (1 + 2) or (3 + 4) and the corresponding cell Nernstian voltage ($V = -\Delta G/4F$), where the symbols have their usual meaning) are identical, however, the individual electrode reactions are distinct, as are the ionic transport mechanisms [13]. The difference in the chemical potential of gases at the electrodes is responsible for the movement of the ionic species. However, the electrochemical reactions occur mainly at electrode/electrolyte interface, within a range of few micrometers into the electrodes from the electrolyte. This interfacial zone is termed as the functional layer while the remaining part of the electrode is primarily a current collector microstructure that should be porous to allow gas access to the functional layer [14].

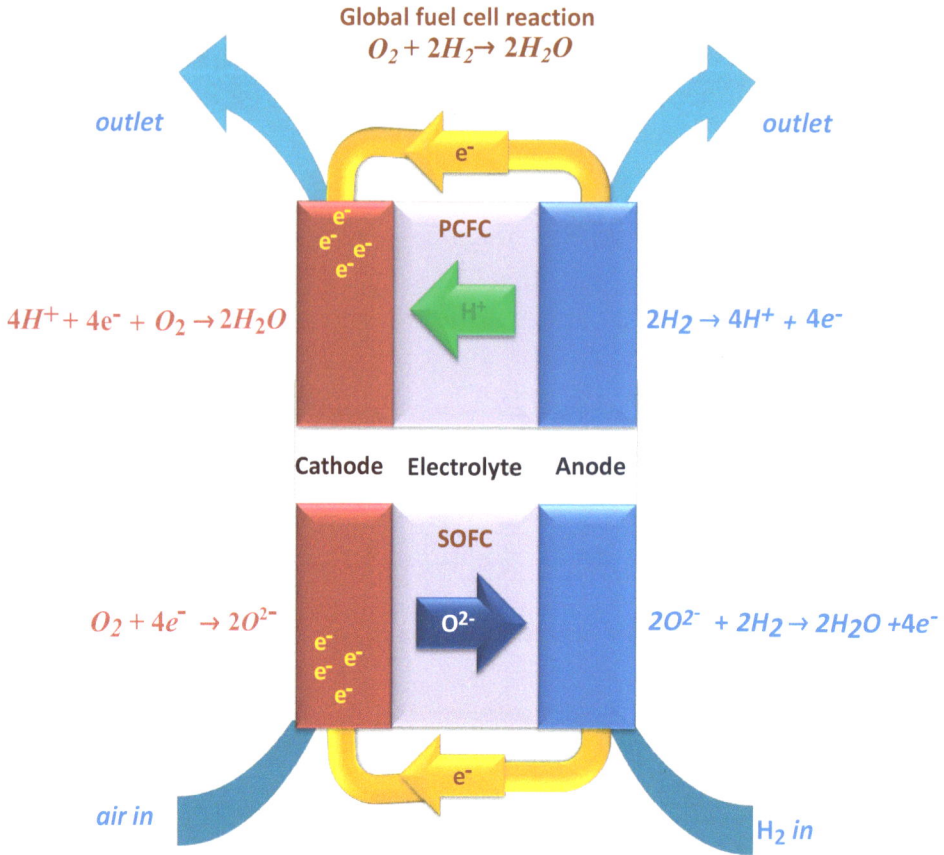

Fig. (2). Schematics of a solid state fuel cell in the case of oxide-ion conductor (SOFC) and proton conductor (PCFC).

The electrolyte must be dense and gas tight to avoid the direct mixing of fuel and oxidant gases. In PCFCs, both the electrolyte and electrodes are based on stable ceramic materials. The current density is obtained under cell operation as long as the fuel and oxidant gases are provided to the cell [14, 15]. An open circuit voltage (OCV) of ~1 volt is attained when the cell is at unloaded state, defined by the Nernst potential [14], which is expressed as

$$E = E^0 - \frac{RT}{2F} ln \left[\frac{pH_2O}{pH_2 \cdot \sqrt{pO_2}} \right] \tag{5}$$

where, E = Nernst potential, R = Universal gas constant, T = Temperature, F = Faraday constant, pH_2, hydrogen partial pressure, pH_2O, water vapour partial pressure, pO_2, oxygen partial pressure. As the maximum voltage of a single cell is close to 1 V, several cells must be connected in series to form a so called fuel cell "stack" to obtain higher voltages for practical application purposes [16]. A stack can be defined as a set of cells; where each cell is separated by an interconnector, whereas, the fuel cell plant is designed based on the desired power output in stacks, connected either in series or parallel [14, 16].

Choice of Materials for PCFCs

A PCFC is composed of two electrodes (the anode and the cathode) and a ceramic proton conducting solid electrolyte. Each and every component of the PCFC must be tailored to their specific function and, thus, must meet certain requirements [16, 17]. These are:

- tolerable chemical and structural stability during the cell operation at high temperatures;
- suitable conductivity (ionic for the electrolyte and electronic or mixed electronic and ionic to function as an electrode);
- proper percolation pathway between metallic and ceramic phase in anodes;
- no inter-diffusion of elements between the cell component materials;
- similar thermal expansion among the cell components, to avoid cracking during the fabrication and cell operation;
- gas tight and dense electrolyte to prevent gas mixing;
- porous electrodes to allow the gas transport near to the reaction sites;
- good mechanical strength;

In addition to the above criteria, the component materials must be of low cost, offer easy fabrication and should not affect the sequential fabrication processing upon addition of further cell components. The ceramic materials used for protonic ceramic fuel cells are described below.

ELECTROLYTES

The most widely studied PCFCs consist of the perovskite family of ABO_3 ceramic oxide structure [3, 6, 18 - 21]. Such structure includes a divalent alkaline earth element, such as Ba^{2+}, Sr^{2+} and Ca^{2+}, in the A-cation site, while a tetravalent element, usually Ce^{4+} or Zr^{4+}, in the B-cation site. The ionic conductivity is assured when the B-site is partially substituted with suitable acceptor elements, such as Y^{3+}, In^{+3} or Gd^{3+} trivalent cations, leading to the formation of charge compensating oxygen vacancies that can be hydrated [3, 18]. This is the case of some of the most well-known compositions, $BaZr_{1-x}Y_xO_{3-\delta}$ (BZY) [17, 22 - 24] and $BaCe_{1-x}Y_xO_{3-\delta}$ (BCY) [25 - 28]. Among these two compositions, the yttrium-doped barium cerate (BCY) performs best in the low temperature range in terms of ionic conductivity, although it reacts with acidic gases, such as CO_2, SO_2 and water vapour [3, 17, 29]. Due to this, the fuel cell applicability of these materials has been delayed. Alternatively, yttrium-doped barium zirconate (BZY) shows good chemical stability under CO_2 containing atmosphere, but its performance in terms of electrical conductivity is lower than that of BCY, with the additional complication of poor process ability, requiring high sintering temperatures and extended annealing times [2, 30, 31].

The introduction of zirconium in barium cerates lattice leads to improvement in processability and chemical stability, albeit with a negative impact on total conductivity (*e.g.* $Ba(Ce,Zr)_{1-y}Y_yO_{3-\delta}$, BCZY) [4, 32, 33]. Therefore, a compromise between these factors is commonly attempted by adjusting the Ce:Zr ratio [32, 34, 35].

Complex-perovskite materials, such as $Ba_3Ca_{1.18}Nb_{1.52}Y_{0.3}O_{9-\delta}$, have also been studied [36], typically showing conductivities (5.3 x 10^{-3} S cm^{-1} in wet air at 600 °C) that are slightly lower than that attainable from the aforementioned simple perovskites. Other ceramic oxide structural families also have demonstrated proton conductivity, such as Ca-doped $LaNbO_4$, a material that changes from monoclinic (Fergusonite-type structure) to tetragonal (Scheelite phase) around 500 °C with proton conductivity in the order of ~ 5×10^{-4} S cm^{-1} at 600 °C [37]. Another example is lanthanum tungstate based phases with fluorite-type structure that are reported to show reasonable levels of proton conduction at intermediate temperatures ~ 1 x 10^{-3} Scm^{-1} at 600 °C [38, 39].

Synthesis

The traditional approach to synthesize proton conducting perovskite materials is by solid-state reaction, intimately mixing the precursor powders, followed by calcination [33, 40, 41]. For example, a common method for making fine, single phase BCZY powder is by solid-state reaction using carbonate and oxide precursors:

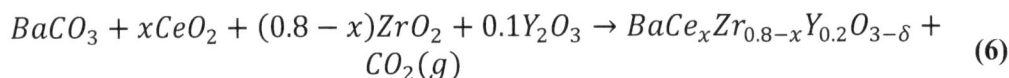

$$BaCO_3 + xCeO_2 + (0.8 - x)ZrO_2 + 0.1Y_2O_3 \rightarrow BaCe_xZr_{0.8-x}Y_{0.2}O_{3-\delta} + CO_2(g) \tag{6}$$

Nonetheless, Sawant *et al.* [42], highlighted that multiple regrinding and refining steps are required by the solid state route to obtain pure perovskite phases of the zirconate and cerate materials of sufficient homogeneity. Several other synthesis procedures have been suggested in the literature for the preparation of

$Ba(Ce,Zr)_{1-y}Y_yO_{3-\delta}$ materials, including glycine–nitrate combustion [32, 34, 42 - 44], sol–gel [32, 34] and pechini methods [42]. Such soft chemical synthesis routes offer more homogeneous, uniform mixing of the precursors at the atomic-level as well as a better stabilization of the oxidation state of the metal precursors in the solution state [42]. However, a disadvantage of many of these chemical synthesis routes is related to the release of NO_x gases during the heating originated by the decomposition of nitrate precursors. Moreover, many of these soft chemical precursors are expensive when compared to simple oxides [32, 34, 42 - 44]. In order to circumvent these issues, a new, environmentally friendly, acetate–H_2O_2 combustion process for the synthesis of multi-element ceramic oxide materials, was developed by Nasani *et al.*, which can be applied to the current BCZY electrolyte materials [4]. This method avoids the use of polluting nitrate precursors by the use of cost effective metal acetates and a hydrogen peroxide solution, to provide the necessary fuel/oxidant mixture for combustion, making this technique both environmentally friendly and cheaper than the nitrate alternative.

In the $BaCe_{0.8-x}Zr_xY_{0.2}O_{3-\delta}$ ($0 \leq x \leq 0.8$) solid solution the symmetry increases with increasing Zr-content from orthorhombic to cubic. Table **1** summarizes the lattice parameters and the unit cell volume for the sintered pellets estimated from the XRD patterns (Fig. **3**) formed by the acetate combustion process [4]. The lattice parameters of all the oxide compositions are in line with published data for other synthesis routes, highlighting the effectiveness of this method [32, 35, 45, 46].

Table 1. Lattice parameter and unit cell volume of $BaCe_{0.8-x}Zr_xY_{0.2}O_{3-\delta}$ materials synthesized by nitrate free acetate–H_2O_2 combustion method [4].

Nomenclature	x (Zr)	Lattice parameters (Å)			Space group	Unit cell volume (Å³)
		a	b	c		
BCY	0	8.921	6.182	6.184	Pnma	341.03
BCZY71	0.1	8.859	6.16	6.162	Pnma	336.3
BCZY44	0.4	4.32	4.32	4.32	Pm-3m	80.61
BCZY26	0.6	4.268	4.268	4.268	Pm-3m	77.74
BZY	0.8	4.222	4.222	4.222	Pm-3m	75.29

The level of distortion of a perovskite lattice can be estimated by the Goldschmidt tolerance factor (t):

$$t = \frac{(R_A + R_O)}{\sqrt{2}(R_B + R_O)} \qquad (7)$$

where R_A and R_B correspond to the ionic radius of the cations occupying the A and B site respectively, while R_O is the oxygen ionic radius. Stable perovskite structures can be obtained when t ranges between 0.75 – 1.0. Cubic symmetry has been observed for values in the range 0.95 – 1.04, while compounds having a tolerance factor in the range 0.75 – 0.90 typically have orthorhombic symmetry [2, 47]. For this reason, the choice of dopants, namely their respective charges and sizes must be carefully considered as these factors can have a strong impact on perovskite stability and crystallographic structure [2, 47].

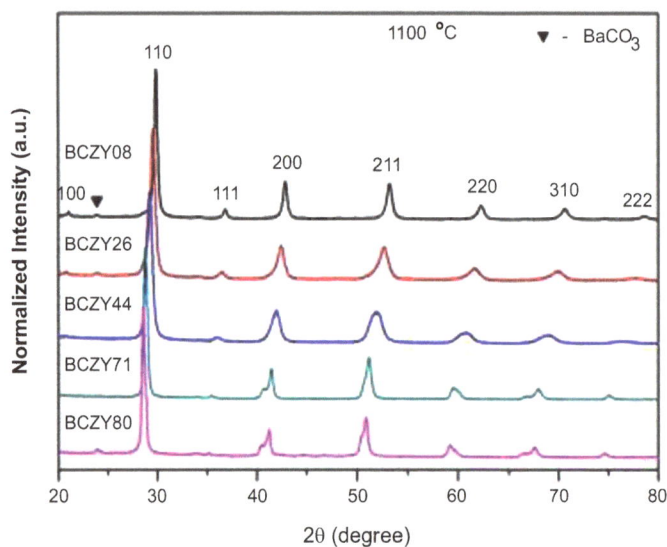

Fig. (3). XRD patterns of $BaCe_{0.8-x}Zr_xY_{0.2}O_{3-\delta}$ (BCZY) (x = 0, 0.1, 0.4, 0.6 and 0.8) powders calcined at 1100 °C, synthesized by nitrate free acetate–H_2O_2 combustion method [4].

Sintering and Microstructure

Some major issues of the zirconate materials are related to their low grain growth and poor densification. Dense zirconium-rich compounds can only be obtained after sintering at very high temperatures >1600 °C [2, 29, 30, 48], temperatures that can lead to unwanted evaporation of barium. Such Ba-losses are reported to have negative impact on resultant levels of proton conductivity [30]. In addition, the total conductivity of these materials is further hindered by poor grain growth that results in ceramics containing a high fraction of resistive grain boundaries [49].

A common strategy to overcome these problems is the use of sintering aids. Babilo *et al.* [50] found that the introduction of transition element dopants, such as ZnO, CuO or NiO (added up to 4 mol%) were effective to promote densification and grain growth in BZY ceramics, with dense materials being obtained under the much milder sintering conditions of 1300 °C for 4 hours. Guo *et al.* [51, 52] studied the densification of BCZY44 using ZnO as a sintering additive by applying three different sintering approaches, and showed that the electrolyte could be densified at 1450 °C. Another example is the work of Tao and Irvine [48] who used ZnO to form the $BaCe_{0.5}Zr_{0.3}Y_{0.16}Zn_{0.04}O_{3-\delta}$, solid solution, increasing sinterability without significantly impairing total conductivity; total conductivity (in wet 5% H_2) exceeded 10 mS cm^{-1} above 600 °C.

However, in the majority of cases these sintering aids have a negative impact on the bulk BZY proton conductivity. Therefore, solving the BZY densification problem, without using sintering aids is still an ongoing goal for PCFC development and, thus, the study of high temperature proton conducting electrolytes remains at a fundamental stage for use in real applications.

As an example of the BCZY system without the use of any sintering additive, Fig. (**4**) shows the pristine surface of BCZY ceramics prepared by acetate combustion and densified at 1500 °C. The micrographs show the low cerium-content, BCZY26, material to have poor densification. Nonetheless, the more cerium rich BCZY80, BCZY71 and BCZY44 pellets are shown to be fully dense (description of nomenclature can be found in Table **1**. The corresponding relative densities and average grain sizes are shown in Table **2**. The average grain size decreases substantially with introduction of Zr in the BCY lattice [32 - 34, 42].

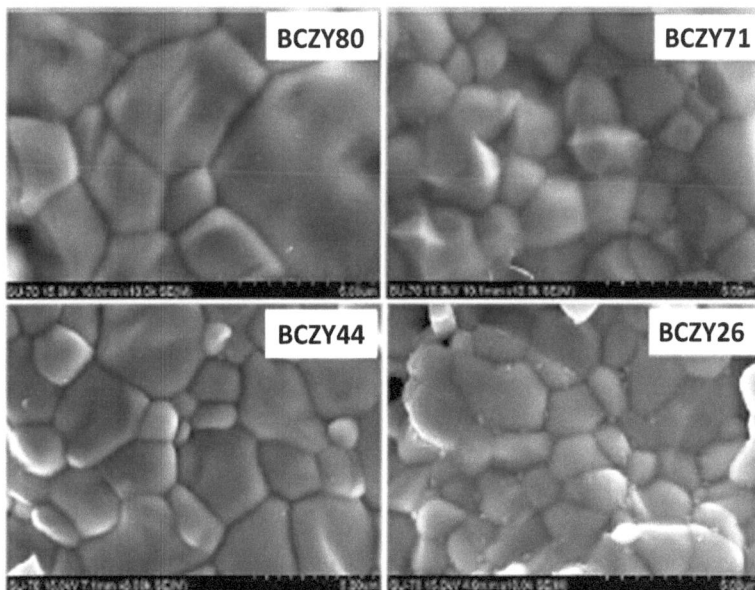

Fig. (4). Scanning electron micrographs of surface of BCZY pellets sintered at 1500 °C for 8 h [4].

Table 2. Grain size and relative density of $BaCe_{0.8-x}Zr_xY_{0.2}O_{3-\delta}$ materials synthesized by nitrate free acetate–H_2O_2 combustion method, isostatically pressed at 200 MPa and sintered at 1500 °C for 8 h [4].

x (Zr)	Relative Density (%)	Grain Size (μm)
0	98.5	4.73 ± 1.65
0.1	98	1.94 ± 0.63
0.4	96	2.61 ± 0.82
0.6	92	1.75 ± 0.51
0.8	82	0.97 ± 0.18

Conductivity

Nasani *et al.* [4] studied the electrical properties (total, bulk and grain boundary) conductivities for BCZY71, BCZY44 and BCZY26 materials, (Fig. **5** and Fig. **6**).

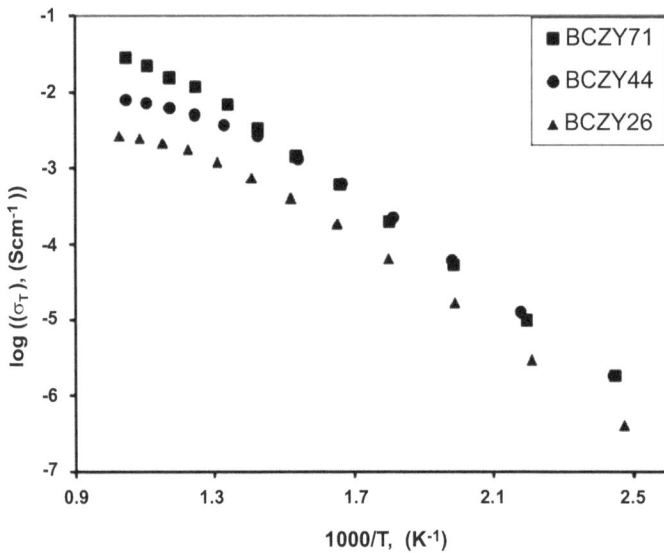

Fig. (5). The temperature dependence of the total conductivity for $BaCe_{0.8-x}Zr_xY_{0.2}O_{3-\delta}$ (x = 0.1, 0.4 and 0.6) materials under wet N_2 atmosphere, pH_2O = 0.026 atm [4].

All studied ceramics have relative densities ≥92% that of the theoretical after sintering the acetate combustion powders at 1500°C for 5 hours [4]. The total conductivity is notably higher for the BCZY71 and BCZY44 compositions with minor contents of Zr. The bulk behaviour is shown in Fig. (**6a**) and shows a decrease with increasing Zr-content (more apparent with increasing temperature) in agreement with further literature data [31, 33, 41, 42, 53].

Concerning the grain boundary response, comparison of the intrinsic grain-boundary properties, can be made by assuming that the proportionality factor of the grain-boundary thickness (*d*) remains constant in the brick-layer model [4]. By such assumption, the grain boundary conductivity can be normalized for microstructural variations by plotting $\sigma_{gb}^{*}=\sigma_{gb}/D$, (Fig. **6b**), where *D* represents the mean grain size measured by SEM analysis (Table **2**).

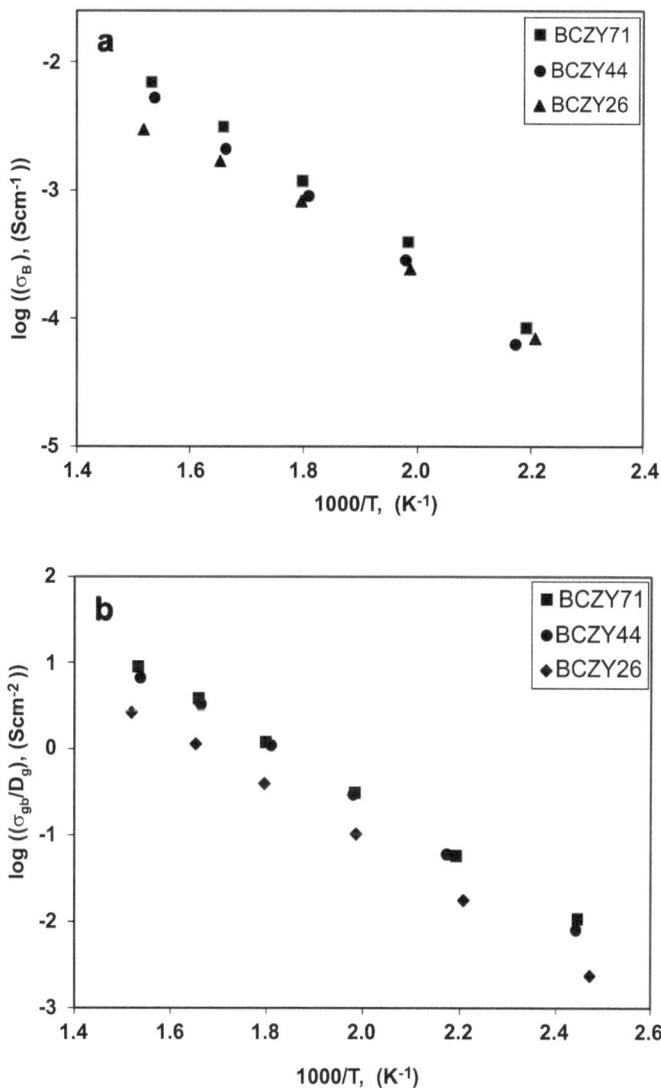

Fig. (6). The temperature dependence of (**a**) the bulk and (**b**) the grain boundary conductivity for $BaCe_{0.8-x}Zr_{x}Y_{0.2}O_{3-\delta}$ (x = 0.1, 0.4 and 0.6) materials under a wet N_2 atmosphere, $pH_2O = 0.026$ atm [4].

Based on these results it is possible to state that the Ce-rich compositions, namely BCZY71 and BCZY44, have similar inherent grain boundary behaviour, while σ_{gb}^{*} is significantly lower for the Zr-rich composition BCZY26. Minor decreases in grain boundary conductivity with increasing Zr-content for cerium rich compositions have also been reported in the works of Yamazaki *et al.* [31] and Ricote *et al.* [33]. One can, thus, conclude that the significantly depleted total conductivity noted for the Zr-rich composition in Fig. (5) results from a combination of poor grain growth (Table 2), low intrinsic grain boundary conductivity and impaired bulk conductivity; factors that can be substantially improved by raising the Ce-content.

Conduction Mechanism

The formation of protonic defects is related to the dissociative adsorption of water at the sample surface, in the presence of oxide-ion vacancies [13]. To promote protonic conductivity, the B-site is, thus, partially substituted with acceptor dopants, such as Y^{3+}, In^{3+}, or Gd^{3+}, to form oxygen vacancies ($V_O^{\cdot\cdot}$) upon charge neutrality. Hydration can then occur through the dissociation of water from the gas phase into a hydroxide ion (OH^-) and a proton (H^+), where the former fills an oxygen vacancy, while the latter forms a hydrogen bond with lattice oxygen. In this way, the reaction generates two protonic defects ($2OH_O^{\cdot}$) [54]. In Kröger-Vink notation, this process is described by the reaction

$$H_2O_{(g)} + V_O^{\cdot\cdot} + O_O^{\times} \rightarrow 2OH_O^{\cdot} \tag{8}$$

Due to the exothermic nature of equation 8, the interaction of oxides with moisture mainly happens in the low-intermediate temperature range, while dehydration occurs at higher temperatures [2, 55, 56]. The formed protonic defects are highly mobile and can migrate by a hopping mechanism from one oxygen atom to the next (Fig. 7).

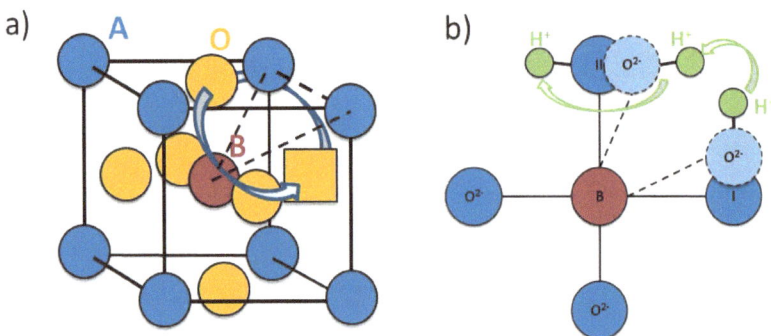

Fig. (7). Schematic view of the types of conduction: (**a**) oxygen ion vacancies in one ABO_3 perovskite involved in the ionic motion of one oxide ion (yellow); and (**b**) a possible proton transfer mechanism from one oxygen ion to another in the same perovskite structure.

ANODES

The anode is another main component of the fuel cell (also called the fuel electrode) and its major role is to oxidise the fuel. The most commonly used anode in SOFCs is that of a composite cermet. A cermet anode consists of both metallic and ceramic (ionic conducting) phases, as proposed by Spacil [57]. This type of composite cermet prevents the coarsening of metallic particles at high temperatures, provides a closer match of thermal expansion coefficient (TEC) with the electrolyte, increases the three phase boundary length (TPBL) between the gas, the ionic and the electronic conducting phases, minimizes gas phase diffusion limitations and reduces polarisation losses. Several metallic composites (Ni, Cu, Pt, Ag, Co, Fe mixed with oxide ion conductors YSZ) have been employed as anodes in solid oxide fuel cells. Among these, nickel cermets have shown promising results due to their high catalytic activity for fuel (H_2, CH_4 *etc.*) oxidation. However, nickel cermets are susceptible to oxidation, carbon deposition and sulphur poising which eventually reduce SOFC performance and longevity. More than a decade of research has been dedicated to optimize and explore the properties of nickel cermets in terms of phase analysis, microstructure, redox stability, cermet composition and novel strategies to avoid degradation issues in oxide-ion conducting SOFCs. Linearly, researchers have adapted similar cermet strategies to PCFCs by employing nickel cermets as anodes based on SOFC anode literature. Hence, PCFC anodes have typically consisted of a metallic (mostly Ni or Cu) phase and an ionically conducting ceramic oxide; in this case a proton conducing oxide, such as $Ba(Ce,Zr)(Y,Yb)O_{3-\delta}$ based phases.

A PCFC anode must be designed to its specific function and, thus, must possess the following basic properties [17, 58]:

- acceptable conductivity (electronic and ionic to function as an electrode, ($>10^2$ S/cm));
- high chemical and structural stability;
- proper percolation pathway between metallic and ceramic phase;
- close thermal expansion coefficient similar among the cell components, to avoid cracking during the cell fabrication and fuel cell operation;
- adequate porosity for gas diffusion to allow unhindered gas transport to the reaction sites;
- good mechanical strength.

A range of compositions have been studied by changing the ratio of Ni to ceramic phase while their structure-property relationships are explained below in detail. With respect to the ceramic phase, anodes that contain $BaCeO_3$-based composites may be unsuitable for real operation in hydrocarbon fuels, owing to the instability of BaCeO$_3$ in steam and CO$_2$ containing atmospheres [29, 59]. On the contrary, acceptor doped BaZrO$_3$-based materials exhibit greater chemical stability in the aforementioned conditions and have, thus, been suggested to be more promising for use as electrodes in these devices [2, 29].

The existence of pores in the anode allows gas diffusion, while the reduced nickel phase provides electronic percolation paths and catalytic properties. The reason behind the selection of a cermet anode is the possibility to raise the density of electrochemically active reaction sites, an important characteristic of anodes in terms of microstructure [60]. However, it remains a very important and open topic to attain basic knowledge on PCFC anodes due to their current infancy. Such work is especially interesting since both nickel and BZY are protonic carriers in PCFC anode cermets, which can extend the three phase boundary length (TPB) over that which would be possible in oxide-ion conducting SOFC anodes where ionic mobility is limited to solely the ceramic phase. The possible kinetic steps involved in fuel (hydrogen) oxidation on nickel particles and a nickel cermet anode with a proton conducting phase are illustrated in Fig. (**8**) [29]. Here, it can be noted that the proton conducting phase can enhance the three phase boundary length (TPB) active sites through the transfer of protons (H$^+$) to electrolyte [29]. Here the overall hydrogen oxidation reaction taking place at the anode can be expressed as:

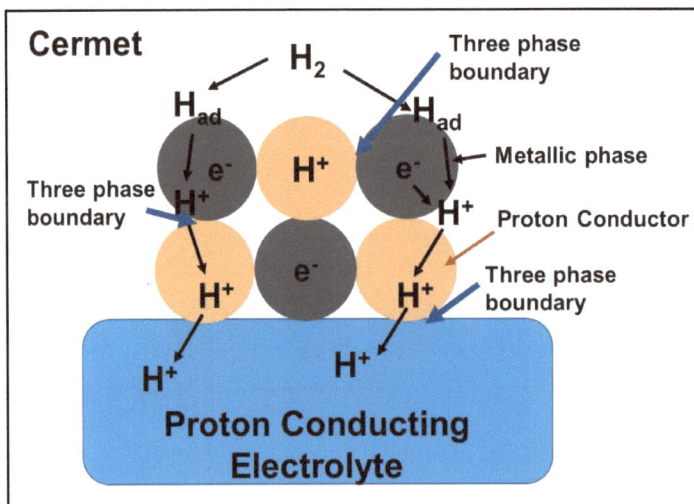

$$H_2 + 2O_O^x \rightarrow 2OH_O^\bullet + 2e^-$$ (9)

Fig. (8). The hydrogen oxidation reaction at anode site nickel cermet-containing proton conducting oxide.

Cermet Compositions

A comprehensive range of nickel cermet compositions (Ni and perovskite proton conducting oxide) have been synthesized and studied, as summarized in Table **3** based on their composition, metal/ceramic ratio and synthesis protocol. In these anodes the nature of the proton conducting phase has been varied, impacting on its bulk proton conductivity, phase purity and stability. So far, Ni-Ca(Zr,Y)O$_{3-\delta}$ and Ni-Sr(Zr,Y)O$_{3-\delta}$ [61], Ni-Ba(Zr,Y)O$_{3-\delta}$ [62, 63], Ni-Ba(Ce,Y)O$_{3-\delta}$ [64 - 67], Ni-Sr(Ce,Yb)O$_{3-\delta}$ [68], Ni-Ba(Ce,Zr,Y,Yb)O$_{3-\delta}$ [69], Ni-Ba(Ce,Zr,Y)O$_{3-\delta}$ [70], NiO-Ba(Ce,In,Y)O$_{3-\delta}$ [71] and NiO-Ba(Ce,Nb,Y)O$_{3-\delta}$ [72] anodes have been proposed for PCFCs. Aside from nickel cermets, only limited other metallic phases have been tested to date, such as Fe-BaCe$_{0.2}$Zr$_{0.6}$Y$_{0.2}$O$_{2.9}$ composite anode by Lapina *et al.* [82]. Due to the complexity of using different compositions of proton conducting oxides in anodes, it is difficult to generalise the acronyms of anodes.

Table 3. PCFC anode compositions and their synthesis methods.

Anode Composition	(x)Metal/(1-x)Ceramic Oxide Ratio	Synthesis Method
Ni-BaZr$_{0.85}$Y$_{0.15}$O$_{3-\delta}$	20,30,40,50 vol% Ni	Acetate combustion [63]
NiO-BaZr$_{0.8}$Y$_{0.2}$O$_{3-\delta}$	40,60,70,80 wt% NiO	Nitrate combustion [62]
Ni-CaZr$_{0.95}$Y$_{0.05}$O$_{2.975}$	33 Vol% Ni	Nitrate Combustion
NiO-BaZr$_{0.8}$Y$_{0.16}$Zn$_{0.04}$O$_{3-\delta}$	60 Wt% NiO	Mechanical mixing [73]
Ni-SrZr$_{0.95}$Y0.05O$_{2.975}$	33 Vol% Ni	Nitrate Combustion [74]
Ni-SrCe$_{0.9}$Yb$_{0.1}$O$_{3-\delta}$	33 Vol% Ni	Nitrate Combustion [75]
NiO-BaCe$_{0.7}$Zr$_{0.1}$Y$_{0.2}$O$_{3-\delta}$	45 wt% NiO	Nitrate combustion [76]
Ni-BaCe$_{0.9}$Y$_{0.1}$O$_{2.95}$	35, 45 Vol% Ni	Mixing [65]
Ni-BaCe$_{0.9}$Y$_{0.1}$O$_{2.95}$	55 Vol% Ni	Wet chemical [64]
Ni-BaCe$_{0.9}$Y$_{0.1}$O$_{2.95}$	40,50,60 wt% Ni	Wet chemical [67]
NiO-BaCe$_{0.2}$Zr$_{0.6}$Y$_{0.2}$O$_{3-\delta}$	68 wt% NiO	Solid state reactive sintering [77]
Ni-BaZr$_{0.1}$Ce$_{0.7}$Y$_{0.1}$Yb$_{0.1}$O$_{3-\delta}$	50 vol% NiO	Mixing [69]
NiO-BaZr$_{0.4}$Ce$_{0.4}$Y$_{0.2}$O$_{3-\delta}$	60 Wt% NiO	Ball milling [78]
NiO-BaZr$_{0.4}$Ce$_{0.4}$Y$_{0.2}$O$_{3-\delta}$	70 Wt% NiO	Nitrate combustion [79]
Ni-BaCe$_{0.8}$Y$_{0.2}$O$_{3-\delta}$	19,26,35,45,56 vol%Ni	Mixing [80]
Ni-BaZr$_{0.85}$Yb$_{0.15}$O$_{3-\delta}$	40 vol% Ni	liquid condensation process [81]
NiO-BaCe$_{0.9-x}$In$_x$Y$_{0.1}$O$_{3-\delta}$	50 wt% Ni	Wet chemical [71]
Fe-BaCe$_{0.2}$Zr$_{0.6}$Y$_{0.2}$O$_{2.9}$	5, 20 wt%	Infiltration [82]
NiO-BaCe$_{0.85-x}$Nb$_x$Y$_{0.15}$O$_{3-\delta}$	60 wt% Ni	Wet chemical [72]
Ni-BaCe$_{0.5}$Zr$_{0.3}$M$_{0.2}$O$_3$ (M=Er, Sm)	50 wt% Ni	Nitrate combustion [83]

Synthesis, Phase Analysis and Microstructure

Anodes can be synthesized *via* existing methods such as solid state method, combustion route, wet chemical synthesis and sol-gel process, as summarized in Table **3**. Mainly, anodes have been prepared in two methods: a) in one step, by mixing all individual ceramic oxide compositional elements and metal precursors in a proper metal to ceramic ratio or b) pre-preparing the perovskite ceramic oxide phase first by aforementioned methods and then mixing with metal/metallic oxides to form the composite anode. Nonetheless, the one-step solid state method is likely to suffer from the aforementioned homogeneity problem, which normally requires several grinding and firing steps to avoid unwanted secondary phases.

Among the methods for preparation of PCFC anodes, method b), using wet chemical microwave assisted combustion to evenly distribute the metallic nickel cermet phase has shown very promising results. Nonetheless, the traditional nitrate combustion method, involving nitrate precursors, has been shown to be unsuitable for PCFC anode preparation *e.g.* Ni-BZY anode due to the formation of an acidic solution in which the pre-prepared proton conducting BZY phase is partially decomposed into barium nitrate ($Ba(NO_3)_2$) before combustion, leading to a nonstoichiometric BZY phase and segregated yttria. After combustion, the $Ba(NO_3)_2$ impurity is converted into barium carbonate ($BaCO_3$) and upon further calcination these impurities react with nickel to form an impure barium yttrium nickelate (BaY_2NiO_5) secondary phase along with the nickel oxide and perovskite phases [63]. In order to prevent the BZY phase from its partial decomposition, Nasani *et al.* [63] proposed a new combustion process using metal acetate precursors. This new acetate combustion route offered a neutral precursor solution, thus, preventing degradation of the BZY phase. In the acetate process, metal acetate is considered as the fuel and hydrogen peroxide (H_2O_2) as the oxidant in contrast to the nitrate combustion method where the metal nitrates are considered as oxidants and urea or glycine as a fuel. In both cases the fuel to oxidant ratio (f/o) was fixed to unity according to propellant chemistry.

The unwanted impurity BaY_2NiO_5 phase was successfully avoided by using this new acetate combustion route and phase pure Ni-BZY anodes were formed, as shown in Fig. (**9**). Hence, this study has shown the high importance of the selection of processing route in PCFC anodes to ensure phase purity. A similar impurity phase, BaY_2NiO_5, has also been observed in one step based synthesis methods in the preparation of Ni-Ba(Ce,Zr,Y)$O_{3-\delta}$ anodes [62, 63, 67, 83].

The microstructure and porosity of anodes plays an important role on three phase boundary length (TPBL-protons, electrons and gas phase) and the transfer of ionic species to the electrolyte. In the case of SOFCs the importance of TPBL has been well studied and the anode microstructure has been optimized for the anode reaction. Nonetheless, a significant difference exists between the anode reaction of a SOFC and that of a PCFC. In a PCFC only hydrogen is present at the anode side. In contrast, in a SOFC both hydrogen and product water exist in the anode compartment. Coors *et al.* studied the PCFC anode composition of 68 wt% Ni-BCZY26 without pore former, and showed that the unique microstructure formed by solely reduction of NiO to Ni metal in the anode is sufficient for PCFC functionality, since the proton is only diffusing ion. When the NiO is reduced to Ni in the anode the volume reduction is about 40% with yield of 26% porosity when no pore former is used for an anode of composition 40 vol%Ni. For such samples, the pore size distribution in reduced PCFC anodes measured was shown to be strongly bimodal by mercury intrusion porosimetry. The nanometric pore sizes correspond to reduced nickel metal with a pocketed appearance with a high surface to volume ratio, whereas, the large pores relate to the network created between Ni and the proton conducting phase from volume contraction of NiO to Ni [77, 84].

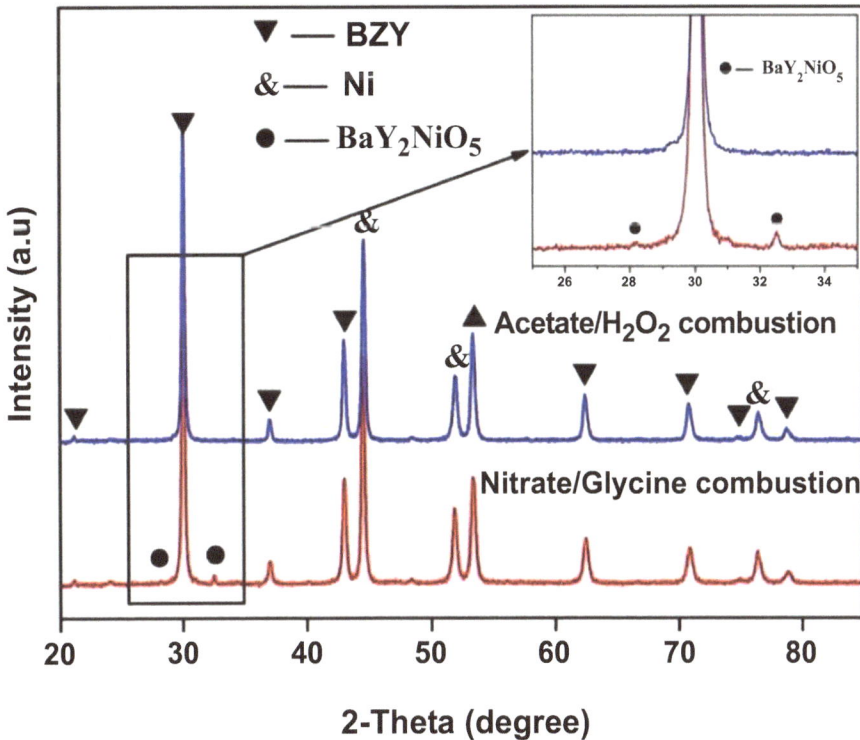

Fig. (9). XRD patterns of Ni-BZY composite powders (reduced in 10% H_2/N_2 gas mixture at 700 °C for 10h, prepared by both nitrate-glycine and acetate-H_2O_2 combustion method).

The microstructure of phase pure 40 vol% Ni-BZY anodes were also extensively explored recently by Nasani *et al.* [63, 84 - 86] A typical PCFC anode microstructure (Fig. **10a**) was formed after sintering at 1400 °C. Fig. (**10a**) demonstrates evident percolation pathways between NiO and BZY phases; bulky and small grains respectively correspond to NiO and BZY particles. After reduction, the 40 vol% contraction of anode can be clearly witnessed in the microstructure offered in Fig. (**10b**). The Ni particles have separated from the surrounding proton conducting phase, leaving narrow gap channels between two phases and nanometric pores within the nickel particles [63]. This microstructure is typical of reported PCFC anodes and has been observed in various anodes *e.g.* Ni-BCZYYb by Rainwater [69], Ni–BaCe$_{0.9}$Y$_{0.1}$O$_{2.95}$ by Laure [64], Taillades [66], Essoumhi [65], Zunic [67], NiO–BaCe$_{0.9-x}$In$_x$Y$_{0.1}$O$_{3-\delta}$ by Zunic [71] and NiO–BaCe$_{0.85-x}$Nb$_x$Y$_{0.15}$O$_{3-\delta}$ by Zunic [72].

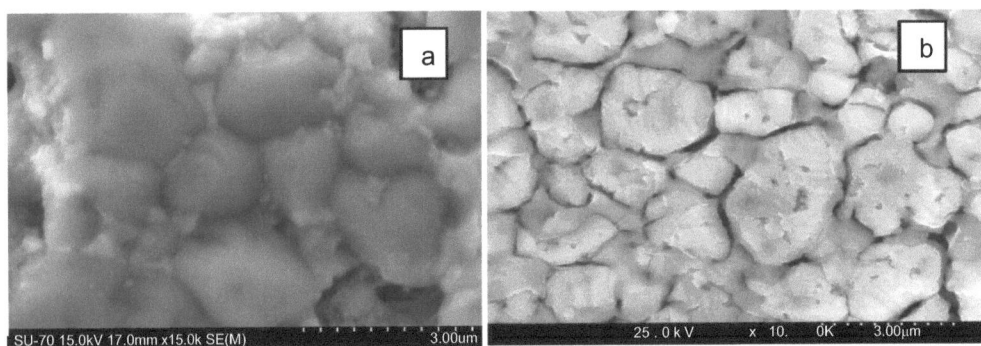

Fig. (10). Scanning electron micrographs of 40 vol% Ni–BZY anodes (**a**) before and (**b**) after reduction [63].

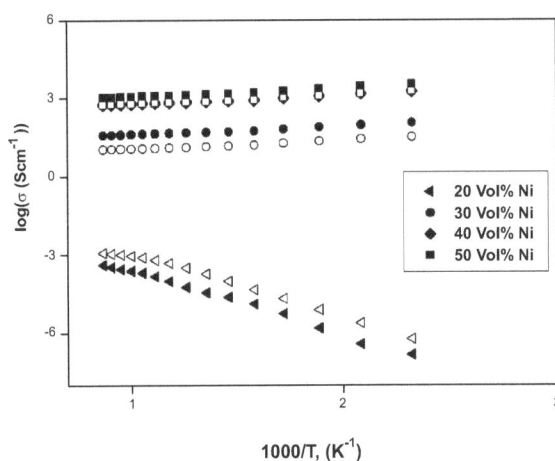

Fig. (11). Arrhenius plots of Ni-BZY anodes under wet (open symbols) and dry (filled symbols) reducing atmospheres [63].

Electrical Conductivity and Effect of Porosity

Another important characteristic of the anode is to provide electrical conductivity. The total electrical conductivity of PCFC anode *e.g.* Ni-BZY anode has a strong dependence on the Ni content in the composition, as depicted in Fig. (**11**). The anode with 20 vol% Ni shows a typical Arrhenius behaviour in wet reducing atmosphere with an activation energy of ~0.58 eV, suggesting that mostly protons are involved in the total conductivity. On further increases of Ni volume to $\geq$ 30%, a typical metallic behaviour is observed. A clear percolation threshold is achieved at above 40 vol% as also suggested by the observed microstructural image shown in Fig. (**10**).

Electrochemical and Redox Behaviour of PCFC Anodes

Bi *et al.* studied composite $NiO-BaZr_{0.8}Y_{0.2}O_{3-\delta}$ anodes with varying weight ratios of nickel to BZY and found the composition 50:50 wt% NiO-BZY to show the lowest polarization resistance [62]. Nasani *et al.* explored the origin of electrode phenomena in (40 vol% Ni) Ni-BZY anodes by studying their electrochemical behavior under typical PCFC operating conditions. The electrochemical impedance spectrum for electrode polarisation can be separated into two main arcs R2, at higher frequency, and R3, at lower frequency at 600 °C, (Fig. **12**). The electrochemical impedance results also further demonstrate that these PCFC anodes show only partial dependence of polarization resistance on pH_2O, but, in contrast, high sensitivity to changes in pH_2. The strong pH_2 dependence originates from the lower frequency term, R3, that demonstrates negative unity dependence on pH_2. This factor was suggested to be related to the dissociative adsorption of H_2 on the anode surface [84]. On the contrary, the higher frequency polarization response, R2, is suggested be related to the migration of protons across the electrode/electrolyte interface at the TPB, with low atmosphere dependence but with significant dependence on microstructure [63, 84, 85], a result that mirrors that observed in oxide-ion conducting SOFC anodes.

A different PCFC composite anode $Ni-BaCe_{0.9}Y_{0.1}O_{2.95}$ of 40, 50 and 60 wt% Ni (36, 45 and 55 vol% Ni, respectively) was studied by Zunic *et al.*, offering clear percolation pathways between Ni and BCY [67], providing metallic behaviour upon reduction. Of these compositions, the authors proposed that the 40 wt% nickel anode was most suitable for PCFCs due to offering the lowest polarisation resistance. Essoumhi *et al.* studied the conductivity and electrochemical behaviour of a Ni-BCY cermet anode containing 35 and 45 vol% Ni using symmetrical cells and their findings concluded that the cermet containing 45 vol% Ni showed the lowest ASR value of 0.4 Ω cm^2 at 600 °C [87].

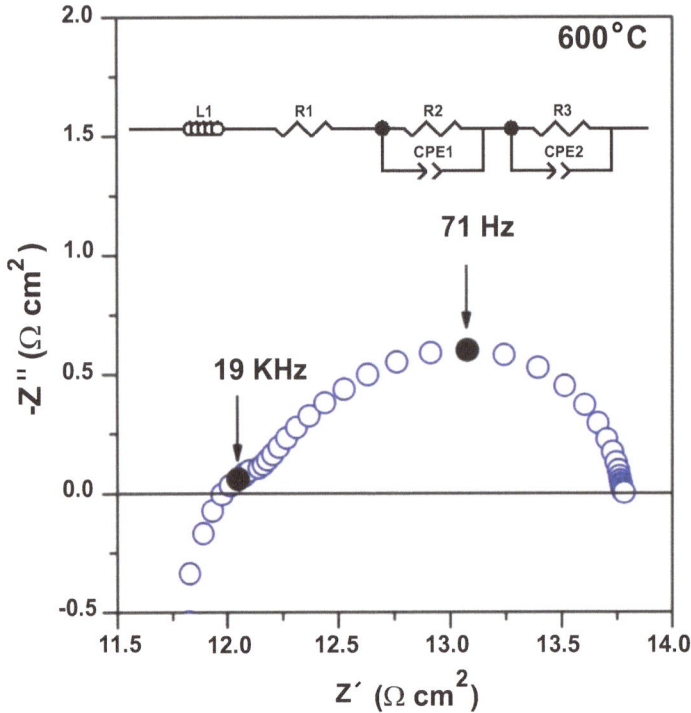

Fig. (12). The electrochemical impedance spectrum (EIS) of a symmetrical button cell with 40 vol% Ni-BZY electrode, measured in wet 10% H_2/N_2 gas mixture at 600 °C [86].

Recently, Rainwater *et al.* reported a decrease in PCFC cell performance on increasing the anode porosity level in Ni-$BaZr_{0.1}Ce_{0.7}Y_{0.1}Yb_{0.1}O_{3-\delta}$ (Ni-BZCYYb) anode supported based BZCYYb electrolyte cells with a $La_{0.6}Sr_{0.4}Co_{0.2}Fe_{0.8}O_{3-\delta}$ (LSCF) cathode [69]. The effect of anode porosity and atmosphere on the polarisation behaviour was further studied in PCFC Ni-BZY (40 vol% Ni) anodes by Nasani *et al.* [84] The PCFC cermet anode with 34% porosity (formed without a pore former) showed lower polarisation resistance (Rp) values than anodes of higher porosity fabricated using pore formers. The concurring trends of Rainwater *et al.* and Nasani *et al.*, therefore, suggest a dependence on porosity that is contrary to that of the oxide-ion conducting SOFC anodes, where the required porosity for optimal performance is much greater; in the region of 50%. The available literature, therefore, shows that both porosity and anode cermet matrix composition, for any fixed metal/ceramic ratio, are highly important factors to be considered with respect to their impact on anode performance.

A PCFC composite anode containing a different metallic phase, Fe-$BaCe_{0.2}Zr_{0.6}Y_{0.2}O_{2.9}$, was produced through an infiltration method but was shown to be degraded during thermal cycles with loss of electronic percolation and coarsening of the iron particles [82].

The involvement of the proton conducting ceramic oxide phase in cermet anodes was investigated by comparing an anode with a proton-conducting ceramic-phase Ni-BZY to an anode with a non-proton conducting phase Ni-BaZrO$_3$ (Ni-BZO). The impedance spectrum of Ni-BZO showed three arcs at high frequency R2, intermediate frequency R3 and low frequency R4. The additional arc R4 was considered to be related to the diffusion polarization as its magnitude decreased significantly in 10%H$_2$/He atmospheres in comparison to that of 10%H$_2$/N$_2$ due to the significantly lower diffusion volume of He than N$_2$. Moreover, the R2 and Rp, values of Ni-BZY were shown to be lower than that of Ni-BZO, under standard operating conditions, whereas for R3 this trend was reversed [85]. The study highlighted that the R2 term was strongly related to the level of proton conductivity in the ceramic oxide phase of the cermet anode and this can be highly important in determining the total polarisation resistance Rp in typical PCFC operating conditions. The result of Nasani *et al.* was in agreement with an earlier study performed on Ni-CaZr$_{0.95}$Y$_{0.05}$O$_{2.975}$ (Ni-CZY) and Ni-SrZr$_{0.95}$Y$_{0.05}$O$_{2.975}$ (Ni-SZY) anodes for PCFCs, prepared by identical methods, that showed the lower total polarization resistance to be provided by the Ni-SZY cermet anode, suggested to be related to the higher proton conductivity of SrZr$_{0.95}$Y$_{0.05}$O$_{2.975}$ with respect to CaZr$_{0.95}$Y$_{0.05}$O$_{2.975}$ [61]. The above studies reveal that proton conducting oxide phase plays an important role in PCFC anodes to increase the three phase boundary length and active reaction sites.

The nickel cermet anodes are easily prone to oxidation when exposed to air and unstable due to the bulk volume expansion of Ni upon its re-oxidation [88]. In real time fuel cell test conditions, the cell may undergo several oxidation and reduction cycles (*i.e.* redox cycle) due to the interruption of fuel supply, power failure, leakage of seal *etc*. To minimize or avoid the problems with nickel, and improve the robustness of the anode, a number of alternative materials have therefore, also been tested as electrodes in oxide-ion conducting SOFCs [29, 89 - 91]. The main groups are nickel alloys with more oxidation resistant metals, such as Cu. However, the composite alloy anode materials have been shown to be catalytically less effective than nickel and are also microstructurally unstable [89, 90]. Hence, it is also very important to understand the redox behaviour of PCFC nickel anodes. The *in-situ* redox cycling behaviour of Ni-BZY anode was probed by Nasani *et al.* [86] using environmental scanning electron microscopy (ESEM) under redox conditions, (Fig. **13**). After re-oxidation, the Ni is oxidized to NiO with a mushroom shaped appearance where NiO was shown to have grown out of the ceramic phase backbone, due to the significant volume expansion of NiO upon its re-oxidation [92]. This was shown to lead to structural changes in anode microstructure and cracks in ceramic oxide backbone. The microstructural mechanism of anode cermet degradation upon redox cycling in PCFC anodes is, therefore, shown to be very similar to that noticed in the well-known oxide-ion conducting SOFC anodes [93].

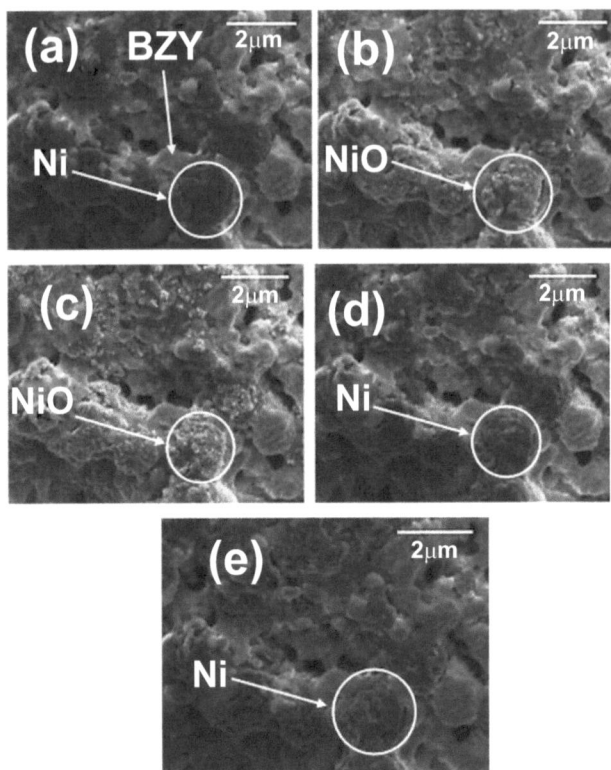

Fig. (13). Environmental Scanning Electron Micrographs (ESEM) of Ni-BZY anode (**a**) first reduced state, (**b**) re-oxidation after 10 min, (**c**) re-oxidised anode after 120 min (Ni volume expansion is clearly visible upon re-oxidation), (**d**) re-reduced after 10 min and (**e**) re-reduced anode after 180 min [86].

The *in-situ* redox cycling behavior of PCFC cermet anodes was also followed by electrochemical impedance spectroscopy. The electrochemical impedance spectra show that both the high frequency (R2) and low frequency (R3) polarisation resistance terms were increased by redox cycling. The ohmic resistance (R_{ohmic}) of the symmetrical cell was also raised due to interfacial delamination of the anode/electrolyte and the presence of micro-cracks, as confirmed by postmortem SEM analysis [86], in agreement with the volume expansion of the Ni-cermet phase upon re-oxidation noted by ESEM examination [86]. Thus, although a high connectivity network provided by a PCFC anode of low porosity is shown to provide preferential performance, this type of microstructure is unfavourable for redox stability.

Another major problem with nickel cermet anodes is carbon mitigation and sulphur poisoning when the cell is operating on carbonaceous fuels, such as methane or biogas [90, 94]. In oxide-ion conducting SOFC this effect has been shown to be able to be suppressed by modifying the anode surface with basic or alkali oxides [95] and a few studies have also stated that nickel anodes that contain barium based mixed or proton conducting perovskite can also tolerate in sulphur and carbon containing atmospheres [96 - 98]. As yet similar mechanistic studies in PCFC remain to be performed.

CATHODES

Only very recently have specifically designed cathodes for the PCFC application been investigated and, in the limited number of papers available, several distinct approaches have been adopted:

i. Mixed oxide-ion-electronic conductors, analogous to classical oxide-ion conducting solid oxide fuel cell (SOFC) cathodes [99, 100];
ii. Single phase mixed proton-electron proton-conducting oxides [101, 102];
iii. Composite electrodes of a proton-conducting oxide with a mixed oxide-ion and electron-conducting component [103].

Although it has been proposed that some proton conductivity in the electrode is required to lower the electrode polarization [104], it remains to be clarified which type of cathode (i) to (iii) is the most efficient and, thus, this remains an important area of study. A PCFC cathode should also offer high stability since it plays major role in cell performance, being at the side of the PCFC where product water is formed [29, 105]. Due to the role of water formation in the PCFC cathode triple conducting cathode materials (conducting protonic, oxide-ion and electronic species) have been suggested to minimize the three phase boundary limitations in PCFCs by allowing all potential conducting species to participate in the electrode kinetics. Thus, as well as electronic conduction and the migration of protons, the triple conducting material is suggested to further spread the electrochemical active sites throughout the structure to enhance the cell performance. In contrast, the existent cathode materials that have been extensively used in oxide-ion conducting SOFCs, such as $La_{1-x}Sr_xMnO_3$ (LSM), $Sm_{0.5}Sr_{0.5}CoO_3$ (SSC) *etc.*, may not be compatible with PCFCs due to limited availability of active sites and the absence of protonic transport, La, Zr and Sr interaction and inter diffusion between electrolyte and cathode materials [103, 106]. Moreover, the oxygen reduction activity of those materials may be less effective under the intermediate temperatures of PCFC operating conditions. Thus, this area remains an essential and open field of ongoing study [29, 103, 105].

The most widely studied cathode materials so far for PCFC applications have been rare earth doped nickelates (R_2NiO_4, R = Pr, Nd, rare earth elements) [100], $Ba(Pr_{1-x}Gd_x)O_{3-\delta}$, $PrBaCuFeO_{5+x}$ and   the   composite   cathodes $La_{0.6}Sr_{0.4}Co_{0.2}Fe_{0.8}O_3/Ba(Zr_{0.1}Ce_{0.7}Y_{0.2})O_3$, $Sm_{0.5}Sr_{0.5}CoO_3/BaCe_{0.8}Sm_{0.2}O_{2.9}$,   and $Ba_{0.5}Sr_{0.5}Co_{0.8}Fe_{0.2}O_3/BaCe_{0.8}Sm_{0.2}O_{2.9}$ [2, 105]. Among all the homogeneous materials, the nickelate, Pr_2NiO_4, cathode has shown the lowest overpotential losses [100, 107, 108]. For the composite cathodes, more compositional flexibility is available and phases offering high electronic conductivity are generally mixed together with well-known proton conducting oxides from the zirconate and cerate perovskite families to increase the cell performance. The applicability of a cathode material for PCFC is dependent not only on its electrochemical performance, but also its thermal expansion coefficient and chemical compatibility with electrolyte material. Platinum has also been used as cathode but is shown to degrade upon long time operation, to exhibit large overpotential losses and is also prohibitively expensive.

The cathode reaction of a PCFC involves the simultaneous reduction of oxygen and formation of water given as [2, 105]:

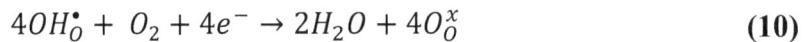

$$4OH_O^{\bullet} + O_2 + 4e^- \rightarrow 2H_2O + 4O_O^x \tag{10}$$

Thus, the cathode may play a key role in the overall performance of PCFCs, due to the formation of water in the cathode side that can result in larger cathode polarization resistances limiting cell performance as temperature is decreased [105, 109].

Single phase mixed triple conducting (H^+, O^{2-}, e^-) cathode materials have also been developed recently, such as $NdBa_{0.5}Sr_{0.5}Co_{1.5}Fe_{0.5}O_{5+\delta}$ that possess a layered perovskite structure showing very promising cell performance that was stable for 500 h at 1023 K [110]. A further new PCFC triple mixed conducting cathode $BaCo_{0.4}Fe_{0.4}Zr_{0.1}Y_{0.1}O_{3-\delta}$ was designed recently by Duan *et al.* The PCFC cells made with this cathode have shown the best performance so far in PCFC history, even at low temperatures <500 °C and maintain performance for 1400 hours without any degradation [98].

CONFLICT OF INTEREST

The authors confirm that they have no conflict of interest to declare for this publication.

ACKNOWLEDGEMENTS

Authors acknowledge financial support from the FCT, FEDER, COMPETE, PTDC/CTM/100412/2008 and PTDC/CTM/105424/2008.

REFERENCES

[1]　Wachsman, E.D.; Marlowe, C.A.; Lee, K.T. Role of solid oxide fuel cells in a balanced energy strategy. *Energy Environ. Sci.,* **2012**, *5,* 5498-5509.
[http://dx.doi.org/10.1039/C1EE02445K]

[2]　Fabbri, E.; Pergolesi, D.; Traversa, E. Materials challenges toward proton-conducting oxide fuel cells: a critical review. *Chem. Soc. Rev.,* **2010**, *39,* 4355-4369.
[http://dx.doi.org/10.1039/b902343g]

[3]　Fabbri, E.; Bi, L.; Pergolesi, D.; Traversa, E. Towards the next generation of solid oxide fuel cells operating below 600 °C with chemically stable proton-conducting electrolytes. *Adv. Mater.,* **2012**, *24,* 195-208.
[http://dx.doi.org/10.1002/adma.201103102]

[4]　Nasani, N.; Dias, P.A.; Saraiva, J.A.; Fagg, D.P. Synthesis and conductivity of $Ba(Ce,Zr,Y)O_{3-\delta}$ electrolytes for PCFCs by new nitrate-free combustion method. *Int. J. Hydrogen Energy,* **2013**, *38,* 8461-8470.
[http://dx.doi.org/10.1016/j.ijhydene.2013.04.078]

[5]　Yamamoto, O. Solid oxide fuel cells: fundamental aspects and prospects. *Electrochim. Acta,* **2000**, *45,* 2423-2435.
[http://dx.doi.org/10.1016/S0013-4686(00)00330-3]

[6]　Brett, D.J.; Atkinson, A.; Brandon, N.P.; Skinner, S.J. Intermediate temperature solid oxide fuel cells. *Chem. Soc. Rev.,* **2008**, *37,* 1568-1578
[http://dx.doi.org/10.1039/b612060c]

[7]　Wachsman, E.D.; Lee, K.T. Lowering the temperature of solid oxide fuel cells. *Science,* **2011**, *334,* 935-939.
[http://dx.doi.org/10.1126/science.1204090]

[8]　Schober, T.; Krug, F.; Schilling, W. Criteria for the application of high temperature proton conductors in SOFCs. *Solid State Ion.,* **1997**, *97,* 369-373.
[http://dx.doi.org/10.1016/S0167-2738(97)00028-3]

[9]　Andújar, J.M.; Segura, F. Fuel cells: History and updating. A walk along two centuries. *Renew. Sustain. Energy Rev.,* **2009**, *13,* 2309-2322.
[http://dx.doi.org/10.1016/j.rser.2009.03.015]

[10]　Zhu, B.; Liu, X.; Zhu, Z.; Ljungberg, R. Solid oxide fuel cell (SOFC) using industrial grade mixed rare-earth oxide electrolytes. *Int. J. Hydrogen Energy,* **2008**, *33,* 3385-3392.
[http://dx.doi.org/10.1016/j.ijhydene.2008.03.065]

[11]　Haile, S.M. Fuel cell materials and components. *Acta Mater.,* **2003**, *51,* 5981-6000.
[http://dx.doi.org/10.1016/j.actamat.2003.08.004]

[12]　Mahato, N.; Banerjee, A.; Gupta, A.; Omar, S.; Balani, K. Progress in material selection for solid oxide fuel cell technology: A review. *Prog. Mater. Sci.,* **2015**, *72,* 141-337.
[http://dx.doi.org/10.1016/j.pmatsci.2015.01.001]

[13] Figueiredo, F.M.; Marques, F.M. Electrolytes for solid oxide fuel cells. *Wiley Interdisciplinary Reviews: Energy and Environment,* **2013**, *2*, 52-72.
[http://dx.doi.org/10.1002/wene.23]

[14] Larminie, J.; Dicks, A.; Larminie, J.; Dicks, A. *Introduction. Fuel Cell Systems Explained*; John Wiley &Sons, Ltd, **2003**, pp. 1-24.
[http://dx.doi.org/10.1002/9781118878330]

[15] de Bruijn, F. The current status of fuel cell technology for mobile and stationary applications. *Green Chem.,* **2005**, *7*, 132-150.
[http://dx.doi.org/10.1039/b415317k]

[16] Singhal, S.C.; Kendall, K. Introduction to SOFCs. In: *High Temperature and Solid Oxide Fuel Cells*; Singhal, S.C.; Kendal, K., Eds.; Elsevier Science: Amsterdam, **2003**; pp. 1-22.
[http://dx.doi.org/10.1016/B978-185617387-2/50018-0]

[17] Stambouli, A.B.; Traversa, E. Solid oxide fuel cells (SOFCs): a review of an environmentally clean and efficient source of energy. *Renew. Sustain. Energy Rev.,* **2002**, *6*, 433-455.
[http://dx.doi.org/10.1016/S1364-0321(02)00014-X]

[18] Malavasi, L.; Fisher, C.A.; Islam, M.S. Oxide-ion and proton conducting electrolyte materials for clean energy applications: structural and mechanistic features. *Chem. Soc. Rev.,* **2010**, *39*, 4370-4387.
[http://dx.doi.org/10.1039/b915141a]

[19] Liu, M.; Lynch, M.E.; Blinn, K.; Alamgir, F.M.; Choi, Y. Rational SOFC material design: new advances and tools. *Mater. Today,* **2011**, *14*, 534-546.
[http://dx.doi.org/10.1016/S1369-7021(11)70279-6]

[20] Tao, Z.; Yan, L.; Qiao, J.; Wang, B.; Zhang, L.; Zhang, J. A review of advanced proton-conducting materials for hydrogen separation. *Prog. Mater. Sci.,* **2015**, *74*, 1-50.
[http://dx.doi.org/10.1016/j.pmatsci.2015.04.002]

[21] Medvedev, D.; Murashkina, A.; Pikalova, E.; Demin, A.; Podias, A.; Tsiakaras, P. $BaCeO_3$: Materials development, properties and application. *Prog. Mater. Sci.,* **2014**, *60*, 72-129.
[http://dx.doi.org/10.1016/j.pmatsci.2013.08.001]

[22] Soares, H.S.; Zhang, X.; Antunes, I.; Frade, J.R.; Mather, G.C.; Fagg, D.P. Effect of phosphorus additions on the sintering and transport properties of proton conducting $BaZr_{0.85}Y_{0.15}O_{3-\delta}$. *J. Solid State Chem.,* **2012**, *191*, 27-32.
[http://dx.doi.org/10.1016/j.jssc.2012.02.053]

[23] Bae, H.; Choi, J.; Kim, K.J.; Park, D.; Choi, G.M. Low-temperature fabrication of protonic ceramic fuel cells with $BaZr_{0.8}Y_{0.2}O_{3-\delta}$ electrolytes coated by aerosol deposition method. *Int. J. Hydrogen Energy,* **2015**, *40*, 2775-2784.
[http://dx.doi.org/10.1016/j.ijhydene.2014.12.046]

[24] Peng, C.; Melnik, J.; Li, J.; Luo, J.; Sanger, A.R.; Chuang, K.T. ZnO-doped $BaZr_{0.85}Y_{0.15}O_{3-\delta}$ proton-conducting electrolytes: Characterization and fabrication of thin films. *J. Power Sources,* **2009**, *190*, 447-452.
[http://dx.doi.org/10.1016/j.jpowsour.2009.01.020]

[25] Dailly, J.; Marrony, M. BCY-based proton conducting ceramic cell: 1000 h of long term testing in fuel cell application. *J. Power Sources,* **2013**, *240*, 323-327.
[http://dx.doi.org/10.1016/j.jpowsour.2013.04.028]

[26] Hibino, T.; Hashimoto, A.; Suzuki, M.; Sano, M. A Solid Oxide Fuel Cell Using Y-Doped $BaCeO_3$ with Pd-Loaded FeO Anode and $Ba_{0.5}Pr_{0.5}CoO_3$ Cathode at Low Temperatures. *J. Electrochem. Soc.,* **2002**, *149*, A1503-A8.
[http://dx.doi.org/10.1149/1.1513983]

[27] Suksamai, W.; Metcalfe, I.S. Measurement of proton and oxide ion fluxes in a working Y-doped BaCeO$_3$ SOFC. *Solid State Ion.,* **2007**, *178*, 627-634.
[http://dx.doi.org/10.1016/j.ssi.2007.02.003]

[28] Tao, Z.; Zhu, Z.; Wang, H.; Liu, W. A stable BaCeO$_3$-based proton conductor for intermediate-temperature solid oxide fuel cells. *J. Power Sources,* **2010**, *195*, 3481-3484.
[http://dx.doi.org/10.1016/j.jpowsour.2009.12.047]

[29] Fabbri, E.; Pergolesi, D.; Traversa, E. Electrode materials: a challenge for the exploitation of protonic solid oxide fuel cells. *Sci. Technol. Adv. Mater.,* **2010**, *11*, 044301.
[http://dx.doi.org/10.1088/1468-6996/11/4/044301]

[30] Magrez, A.; Schober, T. Preparation, sintering, and water incorporation of proton conducting Ba$_{0.99}$Zr$_{0.8}$Y$_{0.2}$O$_{3-\delta}$: comparison between three different synthesis techniques. *Solid State Ion.,* **2004**, *175*, 585-588.
[http://dx.doi.org/10.1016/j.ssi.2004.03.045]

[31] Yamazaki, Y.; Hernandez-Sanchez, R.; Haile, S.M. Cation non-stoichiometry in yttrium-doped barium zirconate: phase behavior, microstructure, and proton conductivity. *J. Mater. Chem.,* **2010**, *20*, 8158-8166.
[http://dx.doi.org/10.1039/c0jm02013c]

[32] Fabbri, E.; D'Epifanio, A.; Di Bartolomeo, E.; Licoccia, S.; Traversa, E. Tailoring the chemical stability of Ba(Ce$_{0.8-x}$Zr$_x$)Y$_{0.2}$O$_{3-\delta}$ protonic conductors for Intermediate Temperature Solid Oxide Fuel Cells (IT-SOFCs). *Solid State Ion.,* **2008**, *179*, 558-564.
[http://dx.doi.org/10.1016/j.ssi.2008.04.002]

[33] Ricote, S.; Bonanos, N.; Manerbino, A.; Coors, W.G. Conductivity study of dense BaCe$_x$Zr$_{(0.9-x)}$Y$_{0.1}$O$_{(3-\delta)}$ prepared by solid state reactive sintering at 1500 °C. *Int. J. Hydrogen Energy,* **2012**, *37*, 7954-7961.
[http://dx.doi.org/10.1016/j.ijhydene.2011.08.118]

[34] Guo, Y.; Lin, Y.; Ran, R.; Shao, Z. Zirconium doping effect on the performance of proton-conducting BaZr$_y$Ce$_{0.8-y}$Y$_{0.2}$O$_{3-\delta}$ ($0.0 \leq y \leq 0.8$) for fuel cell applications. *J. Power Sources,* **2009**, *193*, 400-407.
[http://dx.doi.org/10.1016/j.jpowsour.2009.03.044]

[35] Katahira, K.; Kohchi, Y.; Shimura, T.; Iwahara, H. Protonic conduction in Zr-substituted BaCeO$_3$. *Solid State Ion.,* **2000**, *138*, 91-98.
[http://dx.doi.org/10.1016/S0167-2738(00)00777-3]

[36] Wang, S.; Chen, Y.; Fang, S.; Zhang, L.; Tang, M.; An, K. Novel Chemically Stable Ba$_3$Ca$_{1.18}$Nb$_{1.82-x}$Y$_x$O$_{9-\delta}$ Proton Conductor: Improved Proton Conductivity through Tailored Cation Ordering. *Chem. Mater.,* **2014**, *26*, 2021-2029.
[http://dx.doi.org/10.1021/cm403684b]

[37] Haugsrud, R.; Norby, T. Proton conduction in rare-earth ortho-niobates and ortho-tantalates. *Nat. Mater.,* **2006**, *5*, 193-196.
[http://dx.doi.org/10.1038/nmat1591]

[38] Shimura, T.; Fujimoto, S.; Iwahara, H. Proton conduction in non-perovskite-type oxides at elevated temperatures. *Solid State Ion.,* **2001**, *143*, 117-123.
[http://dx.doi.org/10.1016/S0167-2738(01)00839-6]

[39] Haugsrud, R. Defects and transport properties in Ln$_6$WO$_{12}$ (Ln = La, Nd, Gd, Er). *Solid State Ion.,* **2007**, *178*, 555-560.
[http://dx.doi.org/10.1016/j.ssi.2007.01.004]

[40]   Ricote, S.; Bonanos, N.; Caboche, G. Water vapour solubility and conductivity study of the proton conductor $BaCe_{(0.9-x)}Zr_xY_{0.1}O_{(3-\delta)}$. *Solid State Ion.,* **2009**, *180*, 990-997.
[http://dx.doi.org/10.1016/j.ssi.2009.03.016]

[41]   Ricote, S.; Bonanos, N.; Marco de Lucas, M.C.; Caboche, G. Structural and conductivity study of the proton conductor $BaCe_{(0.9-x)}Zr_xY_{0.1}O_{(3-\delta)}$ at intermediate temperatures. *J. Power Sources,* **2009**, *193*, 189-193.
[http://dx.doi.org/10.1016/j.jpowsour.2008.11.080]

[42]   Sawant, P.; Varma, S.; Wani, B.N.; Bharadwaj, S.R. Synthesis, stability and conductivity of $BaCe_{0.8-x}Zr_xY_{0.2}O_{3-\delta}$ as electrolyte for proton conducting SOFC. *Int. J. Hydrogen Energy,* **2012**, *37*, 3848-3856.
[http://dx.doi.org/10.1016/j.ijhydene.2011.04.106]

[43]   Tu, C-S.; Chien, R.R.; Schmidt, V.H.; Lee, S-C.; Huang, C-C.; Tsai, C-L. Thermal stability of $Ba(Zr_{0.8-x}Ce_xY_{0.2})O_{2.9}$ ceramics in carbon dioxide. *J. Appl. Phys.,* **2009**, *105*, 103504.
[http://dx.doi.org/10.1063/1.3117835]

[44]   Chien, R.R.; Tu, C.S.; Schmidt, V.H.; Lee, S.C.; Huang, C.C. Synthesis and characterization of proton-conducting $Ba(Zr_{0.8-x}Ce_xY_{0.2})O_{2.9}$ ceramics. *Solid State Ion.,* **2010**, *181*, 1251-1257.
[http://dx.doi.org/10.1016/j.ssi.2010.07.024]

[45]   Ma, X.; Dai, J.; Zhang, H.; Reisner, D.E. Protonic conductivity nanostructured ceramic film with improved resistance to carbon dioxide at elevated temperatures. *Surf. Coat. Tech.,* **2005**, *200*, 1252-1258.
[http://dx.doi.org/10.1016/j.surfcoat.2005.07.099]

[46]   Takeuchi, K.; Loong, C.K.; Richardson, J.W., Jr; Guan, J.; Dorris, S.E.; Balachandran, U. The crystal structures and phase transitions in Y-doped $BaCeO_3$: their dependence on Y concentration and hydrogen doping. *Solid State Ion.,* **2000**, *138*, 63-77.
[http://dx.doi.org/10.1016/S0167-2738(00)00771-2]

[47]   Sammells, A.F.; Cook, R.L.; White, J.H.; Osborne, J.J.; MacDuff, R.C. Rational selection of advanced solid electrolytes for intermediate temperature fuel cells. *Solid State Ion.,* **1992**, *52*, 111-123.
[http://dx.doi.org/10.1016/0167-2738(92)90097-9]

[48]   Tao, S.W.; Irvine, J.T. Stable, Easily Sintered Proton- Conducting Oxide Electrolyte for Moderate-Temperature Fuel Cells and Electrolyzers. *Adv. Mater.,* **2006**, *18*, 1581-1584.
[http://dx.doi.org/10.1002/adma.200502098]

[49]   Bi, L.; Fabbri, E.; Sun, Z.; Traversa, E. Sinteractive anodic powders improve densification and electrochemical properties of $BaZr_{0.8}Y_{0.2}O_{3-\delta}$ electrolyte films for anode-supported solid oxide fuel cells. *Energy Environ. Sci.,* **2011**, *4*, 1352.
[http://dx.doi.org/10.1039/c0ee00387e]

[50]   Babilo, P.; Haile, S.M. Enhanced Sintering of Yttrium-Doped Barium Zirconate by Addition of ZnO. *J. Am. Ceram. Soc.,* **2005**, *88*, 2362-2368.
[http://dx.doi.org/10.1111/j.1551-2916.2005.00449.x]

[51]   Guo, Y.; Ran, R.; Shao, Z. Optimizing the modification method of zinc-enhanced sintering of $BaZr_{0.4}Ce_{0.4}Y_{0.2}O_{3-\delta}$-based electrolytes for application in an anode-supported protonic solid oxide fuel cell. *Int. J. Hydrogen Energy,* **2010**, *35*, 5611-5620.
[http://dx.doi.org/10.1016/j.ijhydene.2010.03.039]

[52]   Liu, Y.; Guo, Y.; Ran, R.; Shao, Z. A novel approach for substantially improving the sinterability of $BaZr_{0.4}Ce_{0.4}Y_{0.2}O_{3-\delta}$ electrolyte for fuel cells by impregnating the green membrane with zinc nitrate as a sintering aid. *J. Membr. Sci.,* **2013**, *437*, 189-195.
[http://dx.doi.org/10.1016/j.memsci.2013.03.002]

[53] Barison, S.; Battagliarin, M.; Cavallin, T.; Doubova, L.; Fabrizio, M.; Mortalo, C. High conductivity and chemical stability of $BaCe_{1-x-y}Zr_xY_yO_{3-\delta}$ proton conductors prepared by a sol-gel method. *J. Mater. Chem.,* **2008**, *18*, 5120-5128.
[http://dx.doi.org/10.1039/b808344d]

[54] Kreuer, K.D. PROTON-CONDUCTING OXIDES. *Annu. Rev. Mater. Res.,* **2003**, *33*, 333-359.
[http://dx.doi.org/10.1146/annurev.matsci.33.022802.091825]

[55] Norby, T.; Larring, Y. Concentration and transport of protons in oxides. *Curr. Opin. Solid State Mater. Sci.,* **1997**, *2*, 593-599.
[http://dx.doi.org/10.1016/S1359-0286(97)80051-4]

[56] Iwahara, H.; Uchida, H.; Maeda, N. Studies on solid electrolyte gas cells with high-temperature-type proton conductor and oxide ion conductor. *Solid State Ion.,* **1983**, *11*, 109-115.
[http://dx.doi.org/10.1016/0167-2738(83)90047-4]

[57] Spacil, H.S. inventor; Electrical device including nickel-containing stabilized zirconia electrode. US patent. 3,503,809, 1970.

[58] Singhal, S.C.; Kendall, K. *High Temperature Solid Oxide Fuel Cells: Fundamentals, Design and Applications*; Elsevier, **2003**.

[59] Fabbri, E.; Depifanio, A.; Dibartolomeo, E.; Licoccia, S.; Traversa, E. Tailoring the chemical stability of $Ba(Ce_{0.8-x}Zr_x)Y_{0.2}O_{3-\delta}$ protonic conductors for Intermediate Temperature Solid Oxide Fuel Cells (IT-SOFCs). *Solid State Ion.,* **2008**, *179*, 558-564.

[60] Suzuki, T.; Hasan, Z.; Funahashi, Y.; Yamaguchi, T.; Fujishiro, Y.; Awano, M. Impact of anode microstructure on solid oxide fuel cells. *Science,* **2009**, *325*, 852-855.
[http://dx.doi.org/10.1126/science.1176404]

[61] Mather, G.C.; Figueiredo, F.M.; Jurado, J.R.; Frade, J.R. Synthesis and characterisation of cermet anodes for SOFCs with a proton-conducting ceramic phase. *Solid State Ion.,* **2003**, *162–163*, 115-120.
[http://dx.doi.org/10.1016/S0167-2738(03)00250-9]

[62] Bi, L.; Fabbri, E.; Sun, Z.; Traversa, E. $BaZr_{0.8}Y_{0.2}O_{3-\delta}$-NiO Composite Anodic Powders for Proton-Conducting SOFCs Prepared by a Combustion Method. *J. Electrochem. Soc.,* **2011**, *158*, B797.
[http://dx.doi.org/10.1149/1.3591040]

[63] Narendar, N.; Mather, G.C.; Dias, P.A.; Fagg, D.P. The importance of phase purity in Ni–$BaZr_{0.85}Y_{0.15}O_{3-\delta}$ cermet anodes – novel nitrate-free combustion route and electrochemical study. *RSC Advances,* **2013**, *3*, 859-869.
[http://dx.doi.org/10.1039/C2RA22301E]

[64] Chevallier, L.; Zunic, M.; Esposito, V.; Di Bartolomeo, E.; Traversa, E. A wet-chemical route for the preparation of Ni–$BaCe_{0.9}Y_{0.1}O_{3-\delta}$ cermet anodes for IT-SOFCs. *Solid State Ion.,* **2009**, *180*, 715-720.
[http://dx.doi.org/10.1016/j.ssi.2009.03.005]

[65] Essoumhi, A.; Taillades, G.; Tailladesjacquin, M.; Jones, D.; Roziere, J. Synthesis and characterization of Ni-cermet/proton conducting thin film electrolyte symmetrical assemblies. *Solid State Ion.,* **2008**, *179*, 2155-2159.
[http://dx.doi.org/10.1016/j.ssi.2008.07.025]

[66] Taillades, G.; Batocchi, P.; Essoumhi, A.; Taillades, M.; Jones, D.J.; Rozière, J. Engineering of porosity, microstructure and electrical properties of Ni–$BaCe_{0.9}Y_{0.1}O_{2.95}$ cermet fuel cell electrodes by gelled starch porogen processing. *Microporous Mesoporous Mater.,* **2011**, *145*, 26-31.
[http://dx.doi.org/10.1016/j.micromeso.2011.04.020]

[67]   Zunic, M.; Chevallier, L.; Radojkovic, A.; Brankovic, G.; Brankovic, Z.; Di Bartolomeo, E. Influence of the ratio between Ni and $BaCe_{0.9}Y_{0.1}O_{3-\delta}$ on microstructural and electrical properties of proton conducting Ni–$BaCe_{0.9}Y_{0.1}O_{3-\delta}$ anodes. *J. Alloys Compd.,* **2011**, *509*, 1157-1162.
[http://dx.doi.org/10.1016/j.jallcom.2010.09.144]

[68]   Mather, G.C.; Figueiredo, F.M.; Fagg, D.P.; Norby, T.; Jurado, J.R.; Frade, J.R. Synthesis and characterisation of Ni–$SrCe_{0.9}Yb_{0.1}O_{3-\delta}$ cermet anodes for protonic ceramic fuel cells. *Solid State Ion.,* **2003**, *158*, 333-342.
[http://dx.doi.org/10.1016/S0167-2738(02)00904-9]

[69]   Rainwater, B.H.; Liu, M.; Liu, M. A more efficient anode microstructure for SOFCs based on proton conductors. *Int. J. Hydrogen Energy,* **2012**, *37*, 18342-18348.
[http://dx.doi.org/10.1016/j.ijhydene.2012.09.027]

[70]   Coors, W.G.; Manerbino, A. Characterization of composite cermet with 68wt% NiO and $BaCe_{0.2}Zr_{0.6}Y_{0.2}O_{3-\delta}$. *J. Membr. Sci.,* **2011**, *376*, 50-55.
[http://dx.doi.org/10.1016/j.memsci.2011.03.062]

[71]   Zunic, M.; Brankovic, G.; Foschini, C.R.; Cilense, M.; Longo, E.; Varela, J.A. Influence of the indium concentration on microstructural and electrical properties of proton conducting NiO–$BaCe_{0.9-x}In_xY_{0.1}O_{3-\delta}$ cermet anodes for IT-SOFC application. *J. Alloys Compd.,* **2013**, *563*, 254-260.
[http://dx.doi.org/10.1016/j.jallcom.2013.02.122]

[72]   Žunić, M.; Branković, G.; Basoli, F.; Cilense, M.; Longo, E.; Varela, J.A. Stability, characterization and functionality of proton conducting NiO–$BaCe_{0.85-x}Nb_xY_{0.15}O_{3-\delta}$ cermet anodes for IT-SOFC application. *J. Alloys Compd.,* **2014**, *609*, 7-13.
[http://dx.doi.org/10.1016/j.jallcom.2014.04.175]

[73]   Luisetto, I.; Bartolomeo, E.D.; D'Epifanio, A.; Basoli, F.; Licoccia, S. Internal Methane Reforming High Temperature Proton Conductor (HTPC) Fuel Cells. *ECS Trans.,* **2011**, *35*, 785-795.

[74]   Mather, G. Synthesis and characterisation of cermet anodes for SOFCs with a proton-conducting ceramic phase. *Solid State Ion.,* **2003**, *162-163*, 115-120.
[http://dx.doi.org/10.1016/S0167-2738(03)00250-9]

[75]   Mather, G. Synthesis and characterisation of Ni–$SrCe_{0.9}Yb_{0.1}O_{3-\delta}$ cermet anodes for protonic ceramic fuel cells. *Solid State Ion.,* **2003**, *158*, 333-342.
[http://dx.doi.org/10.1016/S0167-2738(02)00904-9]

[76]   Zhu, Z.; Sun, W.; Yan, L.; Liu, W.; Liu, W. Synthesis and hydrogen permeation of Ni–$Ba(Zr_{0.1}Ce_{0.7}Y_{0.2})O_{3-\delta}$ metal–ceramic asymmetric membranes. *Int. J. Hydrogen Energy,* **2011**, *36*, 6337-6342.
[http://dx.doi.org/10.1016/j.ijhydene.2011.02.029]

[77]   Coors, W.G.; Manerbino, A. Characterization of composite cermet with 68wt.% NiO and $BaCe_{0.2}Zr_{0.6}Y_{0.2}O_{3-\delta}$. *J. Membr. Sci.,* **2011**, *376*, 50-55.
[http://dx.doi.org/10.1016/j.memsci.2011.03.062]

[78]   Liu, Y.; Guo, Y.; Wang, W.; Su, C.; Ran, R.; Wang, H. Study on proton-conducting solid oxide fuel cells with a conventional nickel cermet anode operating on dimethyl ether. *J. Power Sources,* **2011**, *196*, 9246-9253.
[http://dx.doi.org/10.1016/j.jpowsour.2011.07.051]

[79]   Bi, L.; Fabbri, E.; Traversa, E. Effect of anode functional layer on the performance of proton-conducting solid oxide fuel cells (SOFCs). *Electrochem. Commun.,* **2012**, *16*, 37-40.
[http://dx.doi.org/10.1016/j.elecom.2011.12.023]

[80]   Sawant, P.; Varma, S.; Gonal, M.R.; Wani, B.N.; Prakash, D.; Bharadwaj, S.R. Effect of Ni Concentration on Phase Stability, Microstructure and Electrical Properties of $BaCe_{0.8}Y_{0.2}O_{3-\delta}$ - Ni Cermet SOFC Anode and its application in proton conducting ITSOFC. *Electrochim. Acta,* **2014**, *120*, 80-85.
       [http://dx.doi.org/10.1016/j.electacta.2013.12.061]

[81]   Park, Y-E.; Ji, H-I.; Kim, B-K.; Lee, J-H.; Lee, H-W.; Park, J-S. Pore structure improvement in cermet for anode-supported protonic ceramic fuel cells. *Ceram. Int.,* **2013**, *39*, 2581-2587.
       [http://dx.doi.org/10.1016/j.ceramint.2012.09.020]

[82]   Lapina, A.; Chatzichristodoulou, C.; Holtappels, P.; Mogensen, M. Composite $Fe-BaCe_{0.2}Zr_{0.6}Y_{0.2}O_{2.9}$ Anodes for Proton Conductor Fuel Cells. *J. Electrochem. Soc.,* **2014**, *161*, F833-F7.
       [http://dx.doi.org/10.1149/2.017409jes]

[83]   Kang, W.R.; Lee, K.B. Development of rare earth element-doped $Ni–Ba(Ce/Zr)O_3$ cermets for hydrogen-permeable membranes. *J. Ind. Eng. Chem.,* **2015**, *29*, 194-198.
       [http://dx.doi.org/10.1016/j.jiec.2015.04.003]

[84]   Nasani, N.; Ramasamy, D.; Brandão, A.D.; Yaremchenko, A.A.; Fagg, D.P. The impact of porosity, pH_2 and pH_2O on the polarisation resistance of $Ni–BaZr_{0.85}Y_{0.15}O_{3-\delta}$ cermet anodes for Protonic Ceramic Fuel Cells (PCFCs). *Int. J. Hydrogen Energy,* **2014**, *39*, 21231-21241.
       [http://dx.doi.org/10.1016/j.ijhydene.2014.10.093]

[85]   Nasani, N.; Ramasamy, D.; Antunes, I.; Perez, J.; Fagg, D.P. Electrochemical behaviour of Ni-BZO and Ni-BZY cermet anodes for Protonic Ceramic Fuel Cells (PCFCs) – A comparative study. *Electrochim. Acta,* **2015**, *154*, 387-396.
       [http://dx.doi.org/10.1016/j.electacta.2014.12.094]

[86]   Nasani, N.; Wang, Z-J.; Willinger, M.G.; Yaremchenko, A.A.; Fagg, D.P. In-situ redox cycling behaviour of $Ni–BaZr_{0.85}Y_{0.15}O_{3-\delta}$ cermet anodes for Protonic Ceramic Fuel Cells. *Int. J. Hydrogen Energy,* **2014**, *39*, 19780-19788.
       [http://dx.doi.org/10.1016/j.ijhydene.2014.09.136]

[87]   Essoumhi, A.; Taillades, G.; Taillades-jacquin, M.; Jones, D.; Roziere, J. Synthesis and characterization of Ni-cermet/proton conducting thin film electrolyte symmetrical assemblies. *Solid State Ion.,* **2008**, *179*, 2155-2159.
       [http://dx.doi.org/10.1016/j.ssi.2008.07.025]

[88]   Ettler, M.; Timmermann, H.; Malzbender, J.; Weber, A.; Menzler, N.H. Durability of Ni anodes during reoxidation cycles. *J. Power Sources,* **2010**, *195*, 5452-5467.
       [http://dx.doi.org/10.1016/j.jpowsour.2010.03.049]

[89]   Atkinson, A.; Barnett, S.; Gorte, R.J.; Irvine, J.T.; McEvoy, A.J.; Mogensen, M. Advanced anodes for high-temperature fuel cells. *Nat. Mater.,* **2004**, *3*, 17-27.
       [http://dx.doi.org/10.1038/nmat1040]

[90]   Ge, X-M.; Chan, S-H.; Liu, Q-L.; Sun, Q. Solid Oxide Fuel Cell Anode Materials for Direct Hydrocarbon Utilization. *Adv. Energy Mater.,* **2012**, *2*, 1156-1181.
       [http://dx.doi.org/10.1002/aenm.201200342]

[91]   Mohammed Hussain, A.; Høgh, J.V.; Jacobsen, T.; Bonanos, N. Nickel-ceria infiltrated Nb-doped $SrTiO_3$ for low temperature SOFC anodes and analysis on gas diffusion impedance. *Int. J. Hydrogen Energy,* **2012**, *37*, 4309-4318.
       [http://dx.doi.org/10.1016/j.ijhydene.2011.11.087]

[92]   Faes, A.; Hessler-Wyser, A.; Zryd, A. Van herle J. A Review of RedOx Cycling of Solid Oxide Fuel Cells Anode. *Membranes,* **2012**, *2*, 585.
       [http://dx.doi.org/10.3390/membranes2030585]

[93] Klemensø, T.; Chung, C.; Larsen, P.H.; Mogensen, M. The Mechanism Behind Redox Instability of Anodes in High-Temperature SOFCs. *J. Electrochem. Soc.,* **2005**, *152*, A2186-A92. [http://dx.doi.org/10.1149/1.2048228]

[94] Selman, J.R. Materials science. Poison-tolerant fuel cells. *Science,* **2009**, *326*, 52-53. [http://dx.doi.org/10.1126/science.1180820]

[95] Yang, L.; Choi, Y.; Qin, W.; Chen, H.; Blinn, K.; Liu, M. Promotion of water-mediated carbon removal by nanostructured barium oxide/nickel interfaces in solid oxide fuel cells. *Nat. Commun.,* **2011**, *2*, 357-366. [http://dx.doi.org/10.1038/ncomms1359]

[96] Shishkin, M.; Ziegler, T. Coke-Tolerant Ni/BaCe$_{1-x}$Y$_x$O$_{3-\delta}$ Anodes for Solid Oxide Fuel Cells: DFT+U Study. *J. Phys. Chem. C,* **2013**, *117*, 7086-7096. [http://dx.doi.org/10.1021/jp312485q]

[97] Yang, L.; Wang, S.; Blinn, K.; Liu, M.; Liu, Z.; Cheng, Z. Enhanced Sulfur and Coking Tolerance of a Mixed Ion Conductor for SOFCs: BaZr$_{0.1}$Ce$_{0.7}$Y$_{0.2-x}$Yb$_x$O$_{3-\delta}$. *Science,* **2009**, *326*, 126-129. [http://dx.doi.org/10.1126/science.1174811]

[98] Duan, C.; Tong, J.; Shang, M.; Nikodemski, S.; Sanders, M.; Ricote, S. Readily processed protonic ceramic fuel cells with high performance at low temperatures. *Science,* **2015**. [http://dx.doi.org/10.1126/science.aab3987]

[99] Grimaud, A.; Mauvy, F.; Bassat, J.M.; Fourcade, S.; Rocheron, L.; Marrony, M. Hydration Properties and Rate Determining Steps of the Oxygen Reduction Reaction of Perovskite-Related Oxides as H$^+$-SOFC Cathodes. *J. Electrochem. Soc.,* **2012**, *159*, B683-B94. [http://dx.doi.org/10.1149/2.101205jes]

[100] Dailly, J.; Fourcade, S.; Largeteau, A.; Mauvy, F.; Grenier, J.C.; Marrony, M. Perovskite and A$_2$MO$_4$-type oxides as new cathode materials for protonic solid oxide fuel cells. *Electrochim. Acta,* **2010**, *55*, 5847-5853. [http://dx.doi.org/10.1016/j.electacta.2010.05.034]

[101] Fabbri, E.; Markus, I.; Bi, L.; Pergolesi, D.; Traversa, E. Tailoring mixed proton-electronic conductivity of BaZrO$_3$ by Y and Pr co-doping for cathode application in protonic SOFCs. *Solid State Ion.,* **2011**, *202*, 30-35. [http://dx.doi.org/10.1016/j.ssi.2011.08.019]

[102] Mukundan, R.; Davies, P.K.; Worrell, W.L. Electrochemical Characterization of Mixed Conducting Ba(Ce$_{0.8-y}$Pr$_y$Gd$_{0.2}$)O$_{2.9}$ Cathodes. *J. Electrochem. Soc.,* **2001**, *148*, A82-A6. [http://dx.doi.org/10.1149/1.1344520]

[103] Yang, L.; Zuo, C.; Wang, S.; Cheng, Z.; Liu, M. A Novel Composite Cathode for Low-Temperature SOFCs Based on Oxide Proton Conductors. *Adv. Mater.,* **2008**, *20*, 3280-3283. [http://dx.doi.org/10.1002/adma.200702762]

[104] Batocchi, P.; Mauvy, F.; Fourcade, S.; Parco, M. Electrical and electrochemical properties of architectured electrodes based on perovskite and A$_2$MO$_4$-type oxides for Protonic Ceramic Fuel Cell. *Electrochim. Acta,* **2014**, *145*, 1-10. [http://dx.doi.org/10.1016/j.electacta.2014.07.001]

[105] Peng, R.; Wu, T.; Liu, W.; Liu, X.; Meng, G. Cathode processes and materials for solid oxide fuel cells with proton conductors as electrolytes. *J. Mater. Chem.,* **2010**, *20*, 6218-6225. [http://dx.doi.org/10.1039/c0jm00350f]

[106] Holtappels, P.; Vogt, U.; Graule, T. Ceramic Materials for Advanced Solid Oxide Fuel Cells. *Adv. Eng. Mater.,* **2005**, *7*, 292-302. [http://dx.doi.org/10.1002/adem.200500084]

[107] Dailly, J.; Marrony, M.; Taillades, G.; Taillades-Jacquin, M.; Grimaud, A.; Mauvy, F. Evaluation of proton conducting BCY10-based anode supported cells by co-pressing method: Up-scaling, performances and durability. *J. Power Sources,* **2014**, *255*, 302-307.
[http://dx.doi.org/10.1016/j.jpowsour.2013.12.082]

[108] Quarez, E.; Oumellal, Y.; Joubert, O. Optimization of the Lanthanum Tungstate/Pr_2NiO_4 Half Cell for Application in Proton Conducting Solid Oxide Fuel Cells. *Fuel Cells (Weinh.),* **2013**, *13*, 34-41.
[http://dx.doi.org/10.1002/fuce.201200091]

[109] Nasani, N.; Ramasamy, D.; Mikhalev, S.; Kovalevsky, A.V.; Fagg, D.P. Fabrication and electrochemical performance of a stable, anode supported thin $BaCe_{0.4}Zr_{0.4}Y_{0.2}O_{3-\delta}$ electrolyte Protonic Ceramic Fuel Cell. *J. Power Sources,* **2015**, *278*, 582-589.
[http://dx.doi.org/10.1016/j.jpowsour.2014.12.124]

[110] Kim, J.; Sengodan, S.; Kwon, G.; Ding, D.; Shin, J.; Liu, M. Triple-conducting layered perovskites as cathode materials for proton-conducting solid oxide fuel cells. *ChemSusChem,* **2014**, *7*, 2811-2815.
[http://dx.doi.org/10.1002/cssc.201402351]

SUBJECT INDEX

A

Acetic acid 52, 53, 58
Activation energy 17, 29, 30, 32, 44, 47, 60, 149
Active sites 10, 13, 19, 144, 153
  electrochemical 153
Active surface electrode 34
Air electrode 10
Alkaline earth 46, 131
Alkaline fuel cell 89
Alkoxides 78, 82
Ammonia 55
  diluted 55
Amorphous silica 60
Anode catalyst 97
Anode cermet matrix composition 150
Anode compartment 4, 147
Anode composition 145
Anode conductivity 117
Anode/electrolyte interface 6, 26, 27
Anode/electrolyte structure, integrated 97
Anode extension 28
Anode functional layer (AFL) 101, 124, 125
Anode layers 124
Anode materials 29, 37, 151
  composite alloy 151
  nanostructured 29
  preferred 37
Anode 22, 29, 30, 119, 147, 150, 152
  microstructure 22, 29, 30, 147, 152
  performance 150
  porosity 150
  porosity level 150
  reaction 147
  reduction 119
Anodes 4, 6, 26, 27, 28, 31, 32, 34, 37, 101, 119, 126, 143, 144, 145, 146, 147, 149, 150, 151, 152
  based 27
  carbon-resistant 101
  ceria-based 28
  commercial 31
  composite 145, 146, 149, 150
  fabricated 126

one-step 32
re-oxidised 152
re-reduced 152
thick 34
thicker 119
Anode-supported 20, 30, 100, 108, 127
  cell 100
  configuration 127
  design 108
  fuel cell 20
  single cell 30
Anode supporter 95
Anode surface 149, 153
Apatite 42, 43, 48, 53, 57, 61
  doped 57
Apatite powders 53, 60
Apatite structure 44, 45
Area, electrolyte/cathode interface 11
Asymmetric electrolyte microstructure 116
Autoclaves 57, 58

B

Barium strontium cobalt ferrite (BSCF) 19, 118
Barium zirconate 28, 131, 136
BCZY ceramics 139
Behavior, electrochemical 149
Bulk ceramics 21
BZY ceramics 138

C

Calcining temperatures, high 73, 84
Catalytic activity 9, 10, 11, 14, 16, 19, 20, 37
Cathode 3, 4, 5, 6, 9, 11, 13, 16, 22, 89, 100, 108, 133, 153, 154
  composite 19, 20, 154
  perovskite 89
Cathode bulk 10, 17
Cathode electrodes 116
Cathode layer 17, 96, 98, 100, 115, 122
  sintered 100
Cathode performance 11, 77
Cathode polarization resistances 13, 154

Cathode-supported MT-SOFCs 100
  resultant 100
Cell 17, 18, 30, 31, 70, 96, 97, 99, 100, 117,
  118, 121, 125, 134, 135, 149, 152
  electrolyte-supported 117
  symmetrical 149, 152
Cell components 5, 34, 92, 108, 132, 135, 143
Cell operation 27, 134, 135
Cell performance 20, 126, 153, 154
CeO_2-based ceramic materials 99
Ceramic bipolar connector 89
Ceramic bodies 44, 49, 51, 62, 63
  dense 51
Ceramic fuel cells 89, 96, 131
Ceramic hollow fiber precursors 94, 110, 113
Ceramic hollow fiber technology 103
Ceramic materials 70, 71, 89, 111, 134, 135
  advanced 71
  stable 134
Ceramic membranes 88, 89, 112
Ceramic methods 71
Ceramic oxide backbone 152
Ceramic oxide particles 57
Ceramic oxide ratio synthesis method 145
Ceramic oxide structure 135
Ceramic phase 27, 28, 33, 135, 143, 144
  conductive 33
Ceramic phase backbone 151
Ceramic powders 51, 72, 92, 94
  prepared 72
Ceramic processing, ideal 71
Ceramic product 93
Ceramics 3, 26, 45, 49, 50, 52, 59, 60, 62, 99,
  138, 140
  fluorite-type structure 26
  optical transparent 59
  robust 99
  studied 140
Ceramic tubes 92, 103
Cerium gadolinium oxide (CGO) 101, 108,
  109, 113, 114
CGO anode 124
CGO systems 114
Chemical compatibility 9, 10, 27, 154
Chemical composition 29, 131
Chemical reactions 6, 59, 60, 79, 80, 90, 114
Chemical routes 26, 29, 30, 31, 32
Chemical stability 5, 11, 22, 109, 131, 136, 144

Citric acid 7, 27, 54, 56, 75, 76, 79, 80, 82
Coating Ni-YSZ anode 100
Cold isostatic 88, 91, 92
Combustion 70, 75, 77, 137, 146
  microwave-assisted 77
Combustion method 76, 77, 137, 138, 140
  free acetate–H_2O_2 137, 138, 140
Combustion process 75, 76, 146
Combustion reactions 75, 76
Combustion route, acetate 146
Combustion synthesis 71, 75, 76, 131
Commercial powders mixture 30, 31, 32
Complicated electrochemical processes 9
Composite electrodes 153
Co-precipitation method, carbonate 84
Crystal lattice 29, 43, 73, 74
Crystalline phases 44, 52, 54, 55, 61
Crystal structure 9, 11, 12, 13, 16, 20, 44, 45

D

Decorating cathode microstructure 18
Decorating electrode microstructure 16
Degradation, anode cermet 152
Deionized water 55, 56, 76
Dense electrolyte micro tubes 92
Development of low-temperature cathode
  materials 16
Devices, electrochemical 3, 71, 89, 103
Direct methanol fuel cell (DMFC) 132
Dissociative adsorption 35, 36, 142, 149

E

Electrical conductivity 10, 16, 18, 28, 71, 76,
  124, 136, 149
Electrical conductivity relaxation (ECR) 17, 21
Electrical energy 3
Electrical properties 12, 15, 26, 49, 60, 140
Electrocatalytic activity, high 27, 37
Electrochemical activity 28
Electrochemical adsorption 13
Electrochemical behaviour 149
Electrochemical characterization 26
Electrochemical impedance results 149
Electrochemical impedance spectroscopy (EIS)
  17, 150, 152

Electrochemical kinetics 33
Electrochemical performance 6, 21, 27, 28, 31, 32, 107, 110, 114, 154
  showed excellent 21
  superior 21
Electrochemical performance test 125
Electrochemical phenomena 22
Electrochemical processes 33
Electrochemical properties of perovskites 15
Electrochemical reactions 4, 6, 10, 33, 126, 133
Electrochemical site 13
Electrochemical stability test 74
Electrochemical tests 78
Electrode interface 34
Electrode kinetics 153
Electrode materials 6, 16, 84, 114, 131, 132
  active 16
  potential 114
Electrode phenomena 149
Electrode polarisation 149
Electrode polarization 33, 153
Electrode polarization resistances 16
Electrode pores obstruction 27
Electrode porosity 33
Electrode reactions 133
Electrodes 3, 5, 6, 7, 26, 27, 33, 34, 92, 115, 116, 117, 118, 133, 134, 135
  internal 117
  porous 92, 133, 135
Electrodes for intermediate temperature SOFCs 26
Electrode sintering 90
Electrodes materials 26, 114
Electrodes support 126
Electrode structure 33
Electrode surface 13, 35
Electrode thickness 34
Electroless 117
Electroless plating 115, 117
Electroless-plating 115
Electroless plating formula 116
Electroless plating method 117
Electroless plating recipes, optimized 116
Electrolyte 4, 5, 10, 13, 16, 18, 20, 21, 27, 28, 33, 42, 43, 44, 54, 95, 101, 107, 108, 110, 113, 115, 119, 121, 122, 123, 124, 126, 127, 131, 133, 132, 135, 139
  ceramic 33

ceria-based 28, 108
conductive 16, 132
dense 4, 27, 101, 135
excellent 113
fabricated 126
high temperature proton conducting 139
important 43
magnified 123
mixed composite 54
oxide 44
oxide-ion conducting 13, 18, 21, 133
potential 131
prepared 122
pure CGO 124
thick 20
thicker 115
thin 119
thin layer 108, 110, 127
traditional 42
Electrolyte/anode structure, integrated 97
Electrolyte conductivity 35
Electrolyte film 16, 96, 115
  coated YSZ 96
  thin 16
Electrolyte layer 89, 92, 95, 101, 102, 108, 118, 119, 121, 123, 125, 126
  dense 89
  possessed thinner 108
  thick 92, 95, 118, 126
  thinnest 121
Electrolyte layer thicknesses 98, 121
Electrolyte materials 5, 22, 27, 28, 99, 114, 131, 137, 154
  current BCZY 137
Electrolyte membrane 10, 91, 96, 97
  ceramic 91
  prepared dense 96
Electrolyte powder 95
Electrolyte resistivity 35
Electrolyte self-support 108
Electrolyte-supported MT-SOFC 118
Electrolyte suspension 96
Electrolyte system 113, 114
Electrolyte thickness 91, 95, 100, 119, 121, 122, 125, 126
Electron-conducting component 153
Electron conduction 10
Electronic compensation 15

Electronic conduction 15, 20, 43, 153
Electronic conductivity 6, 26, 27, 28, 43, 113, 114, 154
  good 27
  high 6, 26, 154
  low 43, 113
Electronic conductor 10, 11, 13, 27, 109
  pure 10, 11, 13
Electronic percolation 150
Electronic percolation paths 144
Electron orbital 14
Electrons 4, 6, 10, 13, 14, 15, 16, 18, 27, 33, 34, 125, 133, 146
  backscattered 125
  d-orbital 14, 15
  incoming 4
  releasing 133
Electrons flow 33
Electrons occupying hydrogenic orbitals 15
Electron transfer step 34
Electrostatic-assisted ultrasonic spray pyrolysis 31
Energy 73, 74
  chemical 3, 88, 89
Energy efficiency 89, 90
Environmental scanning electron micrographs 152
Environmental scanning electron microscopy (ESEM) 151, 152
Ethylene glycol 54, 56, 79, 80, 82

F

Fabrication of ceramic hollow fiber 110
Fabrication parameters 107, 109, 110, 116
Fabrication processes 37, 94, 109
Fabrication techniques 88, 126
Fluorite structures 42, 43
Fuel cell, single 18
Fuel cell applicability 136
Fuel-cell components 16
Fuel cell conditions, real 74
Fuel cell design 118
Fuel cell development 3
Fuel-cell fabrication conditions 11
Fuel cell operation 143
Fuel cell plant 135

Fuel cells 3, 19, 21, 35, 82, 88, 89, 97, 98, 101, 113, 119, 122, 132, 135, 143
  resultant 97, 101
Fuel cell system 89
Fuel cell technology 43
Fuel electrode 143
Fuel electrode side 4
Fuel oxidation reactions 6, 26, 28

G

Ga-doped ceria (GDC) 27, 28, 29, 30, 31, 32, 33, 43, 98
Gas diffusion 10, 91, 92, 143, 144
Gas-tightness 101, 102, 124, 125
GDC electrolyte 31, 32
Glycine 75, 76, 77, 82, 146
Glycine–nitrate process and mixture of commercial powders 31
Grain boundary conductivity 141, 142

H

High dense ceramics 45
High electrolyte ohmic loss 115
High performance anode 126
High temperature operation 89
High temperature sintering 95
High-temperature sintering 93
Hollow fiber membranes 94, 97, 110, 114
  ceramic 94, 110
Hollow fiber precursors 93, 110, 111, 116
  extrusion of ceramic 110
Hollow fibers 92, 93, 94, 95, 96, 97, 98, 100, 101, 107, 108, 110, 113, 114, 115, 116, 122, 123, 126
  cathode-supported 122
  electrolyte-supported 110, 113
  inorganic 92, 93
  supported 96, 122
Hollow fiber SOFCs 96, 98, 99, 117, 121
Hollow fibre 88, 94, 102, 125
Hollow fibre anode 102
Homogeneous nanopowders 45, 58
Hydrocarbons 101, 133
Hydrogen oxidation reaction 35, 37, 144
Hydrolysis reaction 51, 52

I

Impedance spectrum, electrochemical 149, 150
Inorganic powders 92, 97
Interface 10, 22, 28, 34, 96, 111, 112, 113, 125,
  133, 149
  cathode/electrolyte 10
  electrode/electrolyte 22, 34, 133, 149
  electrode/gas 28
  electrolyte/electrode 28, 125
Intermediate temperature SOFCs 26, 37
Internal coagulants 97, 102, 115, 116, 124
Ionic conductivity 11, 13, 16, 17, 28, 42, 43, 44,
  48, 113, 114, 132, 135, 136
  high 42, 43, 113
Isotope exchange depth profiling (IEDP) 17, 21
IT-SOFC electrolyte 45
IT-SOFC electrolyte application request 63
IT-SOFCs electrolytes 42

L

Lanthanum nickel ferrite (LNF) 109
Lanthanum nitrate 51, 52, 54, 80
Lanthanum silicate apatite (LSA) 42, 44, 45,
  46, 48, 53
Lanthanum silicates 42, 43, 44, 48, 50, 51, 52,
  54, 56, 59
  apatite-type 43, 54, 56
Lanthanum strontium cobalt ferrite (LSCF) 30,
  94, 95, 109, 115, 150
Lanthanum strontium cobaltite (LSC) 77, 109
Lanthanum strontium ferrite (LSF) 84, 85, 109
Lanthanum strontium manganite (LSM) 77, 84,
  85, 109, 114, 115, 153
Lattice parameters 20, 137
Law, electro neutrality 15
Layers 90, 97, 100, 101, 115, 124, 133
  anode support 97
  ceramic 90
  functional 100, 124, 133
  functional anode 101
  outer 124
  porous anode 101
  thin dense skin 97, 115
Low solution stability 78, 80
Low synthesis temperature 82

Low-temperature cathode materials 16
Low-temperature operating cathodes, potential
  10
Low-temperature solid oxide fuel cells 9

M

Maximum power densities 21, 31, 32, 33, 78,
  115, 116, 117, 121, 123, 125
Membrane 10, 112, 113, 114, 116
  electrolyte conducting 10
  electrolyte-supported 114
Metallic behaviour 149
Metal nitrates 76, 146
Methanol fuel cell, direct 132
Microstructure, porous 32, 62
Microstructure of dual-layer 100
Micro tubes 90, 92, 97, 103, 118
  ceramic 92, 103
  prepared electrode 92
Micro-tubular SOFCs 88, 103, 108, 123
Micro-tubular solid oxide fuel cells 88
Minimal electrolytic resistance losses 108
Mixed Ionic-electronic conductor 10, 13, 26
Mixed oxide-ion-electronic conductors 153
Mixed   proton-electron   proton-conducting
  oxides 153
Modified polymeric complexing method 81, 82
Molten carbonate fuel cell (MCFCs) 89, 132,
  133
MT-SOFC development 92, 108
MTSOFCs, anode-supported 119
MT-SOFCs 90, 95, 101, 114, 118, 119, 121,
  126
  anode-supported 95, 101, 119
  electrolyte-supported 95, 114
  electrolyte-supported hollow fiber 119
  performance of 90, 118, 121
  supported 126

N

Neutron powder diffraction (NPD) 20, 44
Nickel anodes 117, 149, 153
  deposited 117
Nickel cermet anodes 144, 151, 153
Nickel cermets 143, 145

Nickel electrodes, pure 27
Ni-doped ceria cermets 28, 29
Ni/electrolyte interface 36
Nitrate combustion 145
Nitric acid 53, 54, 61
Noble-metal-free electrodes 89

O

Ohmic resistance 34, 35, 152
Open circuit voltage (OCVs) 31, 32, 98, 134
Optimization, microstructural 6, 27
Overpotential 33, 34, 35
Oxalate reverse co-precipitation methods 32
Oxidant 21, 96, 97, 100, 117, 118, 123, 146
Oxidant gases 134
Oxide ions 16, 46, 47, 54, 142
Oxide materials, ceramic 137
Oxides 4, 11, 18, 19, 20, 46, 57, 71, 72, 76, 79, 83, 85, 131, 136, 137, 142, 143, 146
  ceramic 131, 136, 143, 146
Oxygen atoms 12, 15, 142
  pπ orbitals of 15
Oxygen content 20
Oxygen diffusion 9, 19, 20, 21
Oxygen ions 4, 10, 13, 14, 16, 17, 36, 43, 44, 142
  mobility of 16
Oxygen reduction 5, 10, 11, 13, 14, 133
Oxygen reduction reaction (ORR) 4, 6, 9, 10, 11, 13, 14, 16, 17, 18, 19, 21, 35
Oxygen species 13
Oxygen surface exchange 9, 13, 17, 20, 21
Oxygen vacancies 13, 15, 16, 20, 21, 43, 142
Oxygen vacancy concentration 17, 18

P

Particle size, synthesized powder 60
PCFC anode cermets 144
PCFC anode compositions 145, 147
PCFC anode microstructure 148
PCFC anode of low porosity 152
PCFC anode preparation 146
PCFC anodes 143, 144, 146, 147, 148, 149, 151, 152
  reduced 147

reported 148
Pechini method 48, 56, 80, 81, 82, 137
Perovskite ABO3-type cathodes 20
Perovskite oxide electrodes 114
Perovskite oxides 11, 12, 14, 15
Perovskite structure 11, 142
Phosphoric acid fuel cell (PAFC) 132
Phosphoric fuel cell 89
Planar fuel cell 90
Polarization resistance 19, 21
  interfacial electrode 19
  low electrode 21
Polarizations, activation 34
Polyesterification reaction 57, 79
Polymer binder 92, 93, 111
Polymeric complexing method 70, 71, 79, 80
Polymeric membrane formation 111, 112
Polymeric precursor method 32, 70, 79
Polymeric resin 79, 80
Porous anodes 4, 26
Porous layers 115, 118
Porous metal electrodes 27
Porous SDC electrolyte backbones 18
Powder agglomeration 75, 78
Powder Manufacturing Method 30, 31, 32
Powders 28, 29, 31, 32, 33, 44, 45, 49, 50, 53, 59, 63, 76, 77, 78, 79, 81, 82, 84, 85, 147
  composite 32, 147
  nanocomposite anode 29
  starting 28, 29
  synthesize apatite-type lanthanum silicate 63
  synthesize cathode 82
  synthesized 45
Power densities 18, 19, 21, 88, 90, 96, 98, 101, 102, 108, 115, 119, 122, 126
  high volumetric 88, 90
  peak 19, 21, 98, 101
Power generator 89, 90
Precursor powders, ceramic 50
Preparation 70, 112
  ceramic membrane 112
  ceramic powders 70
Process of synthesis 42, 50, 51, 74, 80
Process parameters 45, 55
Production, industrial ceramic powders 55
Properties 5, 9, 10, 11, 13, 16, 28, 30, 32, 70, 113, 125, 143
  electrochemical 16, 125

Proton conducting 131, 133, 135, 151
  ceramic 135
Proton conducting phase 144, 145, 147
Proton conductivity 131, 136, 138, 151, 153
Protonic ceramic fuel cells (PCFCs) 131, 133,
    134, 135, 143, 145, 147, 149, 150, 151,
    153, 154
Protonic defects 132, 142

R

Rare earth silicates 47
Real time fuel cell test conditions 151
Redox behaviour of PCFC anodes 149
Redox cycling 152
Relative densities 32, 44, 45, 52, 55, 63, 139,
    140

S

Scanning electron micrographs 139, 148
SDC electrolyte 19
Secondary ion mass spectrometry (SIMS) 17
Single cells YSZ electrolyte 32
Sintering aids 138, 139
Sintering Condition Temperature 63
Sintering process 21, 42, 44, 93, 110, 115, 127
  low-temperature 21
Sintering techniques 92, 102, 103, 114, 115,
    117, 118
Sintering temperature 27, 50, 51, 59, 60, 63, 76,
    97, 100, 118, 122
Sintering temperatures, high 100, 136
Sm-doped ceria (SDC) 18, 21, 43, 54, 84, 95,
    118
SOFC anode literature 143
SOFC anodes 28, 35, 144, 149, 150, 152
  oxide-ion conducting 144, 149, 150, 152
SOFC Cathodes 10, 20, 37
SOFC components 4, 6, 26, 70, 89, 107
SOFC electrode materials 37
SOFC ELECTRODES 5, 33, 34, 37
SOFC electrolyte 4, 43, 75
SOFC electrolyte application 45
SOFCs 18, 19, 21, 22, 88, 91, 98, 143, 151, 153
  electrolyte-supported 114, 115
  low-temperature 21, 22

oxide-ion conducting 143, 151, 153
  proton conducting 18, 19
  tubular 88, 91, 98
SOFCs 29, 43
  electrodes 29
  electrolyte 43
SOFC systems 28
Sol-gel method 51, 78, 79
Sol-gel process 51, 54, 59, 78, 146
Solid electrolyte 4, 42, 43, 45, 131, 133, 135
Solid state 70, 71, 72, 73, 74, 134, 146
  diffusion 72, 73
  fuel cell 134
  method 146
  reaction 70, 71, 72, 73, 74
Solution 56, 75, 84, 111, 137, 139
  clear 56
  polymeric 111
  solid 75, 84, 137, 139
Spark plasma sintering (SPS) 45, 58, 59
Species 34, 153
  electroactive 34
  electronic 153
Spectroscopies, electrochemical impedance 17,
    152
Spinneret, quadruple-orifice 102, 124, 126
Spinning suspension, preparation of 110
Spinning two-layered ceramic hollow fibers 99
Stability, enhancing electrode 30
Stoichiometric amounts 49, 50, 58, 61
Studied fuel cells 4
Synthesis 31, 32, 48, 50, 51, 53, 56, 57, 58, 59,
    61, 74, 76, 136, 137
  high temperature 59
  hydrothermal 57, 58
Synthesis methods 30, 32, 55, 70, 72, 145
  coprecipitation 55
Synthesis of oxide powders 75
Synthesis procedures 45, 136
Synthesize 29, 49, 50
Synthesize LSA 48, 60, 61
Synthesize LSA powders 45, 48
Synthesize LSA precursor powders 55
Systems 35, 43, 44, 46, 47, 54, 78, 108, 112
  ceramic 112
  hexagonal crystal 43, 44

T

Technique 20, 21, 45, 48, 55, 56, 58, 88, 92, 93, 94, 101, 103, 107, 110, 114, 115, 117, 127, 137
  phase inversion 92, 93, 94, 101
  phase inversion and sintering 92, 103
  phase-inversion and sintering 114, 115, 117
Technologies, current fuel cell 132
Theoretical density 63
Thermal expansion coefficient (TEC) 27, 114, 143, 154
Thickness, total anode 125
Thin layer electrode 89
Three-phase boundary 9
Three phase boundary length (TPBL) 143, 144, 146, 147, 151
Total conductivity 16, 17, 136, 138, 139, 140, 149
Triple phase boundaries (TPB) 5, 6, 7, 9, 27, 28, 31, 33, 34, 125, 126, 144, 149
Tubular ceramic support 89
Tubular design 107, 108, 109
Tubular fuel cell 90
Tubular solid oxide fuel cell 107

U

Uninterruptible power supply (UPS) 90

X

X-ray diffraction (XRD) 42, 49, 54, 56, 61, 62

Y

YSZ and CGO systems 114
YSZ cermets 27, 28
YSZ electrolyte of hollow fiber 115
YSZ electrolytes 37, 44, 78, 94, 99, 116
  based 116
Yttria stabilized zirconia (YSZ) 20, 27, 28, 33, 36, 42, 43, 44, 94, 95, 99, 108, 113, 114, 116

Z

Zirconia balls 49, 59
ZrO2 electrolyte membrane 89

www.ingramcontent.com/pod-product-compliance
Lightning Source LLC
Chambersburg PA
CBHW041728210326
41598CB00008B/811